Roger Tabor's
CAT
BEHAVIOUR

The
complete feline
problem
solver

Roger Tabor's
CAT
BEHAVIOUR

DAVID & CHARLES

For Liz, who traced Herodotus' Ailuroi with me from *Zagazig* to *Bubastis*; and in memory of Helena Sanders, befriender of Venice's cats, Mary Wyatt, feeder of the cats of Fitzroy Square, and Roy Robinson, the geneticist of cats

VIDEOS: In addition to his BBC TV *Cats* series and *Understanding Cats* series, the author has made a number of educational videos on cats with Cats on Film/Bowe-Tennant Productions: *Understanding Your Cat*; *Breaking Bad Habits for Cats*; *The Cat Outdoors*; *Complete Cat Care* and *The Mystery of the Cat*.

AUDIO PACKS: Roger Tabor discusses the neutering of feral cats and wildlife predation by cats in the 'Focus on Ferals' seminar audio pack (Alley Cat Allies, U.S.A. 1994).

PICTURE CREDITS: All photographs by Roger Tabor except the following:
Liz Artindale: pages 2–3, 9, 11(inset), 15(top), 16(top rt), 17(top), 18, 19(top & btm left), 25(top), 27, 28, 29, 32(top rt & btm), 33(rt), 38(left), 55, 66–7, 72, 73(top), 76, 77(top), 88, 90(top), 95(top), 118–19, 123, 125(rt), 130, 132, 134, 137(top)
Marc Henrie: pages 1, 10–11, 12, 13(btm), 80(btm)
BBC Natural History Unit: Anup Shah page 20; Miles Barton page 22(top); Lynn M. Stone page 22(btm); Gerry Ellis page 23
FLPA: Michael Callan page 23
Warren Photographic: Jane Burton page 69

ARTWORKS: Eva Melhuish pages 14, 16, 35, & 111

A DAVID & CHARLES BOOK

First published in the UK in 1997

Copyright © Roger Tabor 1997

Roger Tabor has asserted his right to be identified as author of this work in accordance with the Copyright, Designs and Patents Act, 1988.

A catalogue record for this book is available from the British Library.

ISBN 0 7153 0624 3

Printed in the UK by The Bath Press for David & Charles
Brunel House   Newton Abbot   Devon

Contents

COMMON BEHAVIOURAL PROBLEMS

Introduction

The catte is…in youth swyfte, plyante and merry and lepeth and reseth on all thynge that is before him, and is led by a strawe and playeth therwith. And is a righy hevy beast in age, and ful slepy, and lieth slily in wait for myce…and when he taketh a mouse he playeth therwith, and eateth him after the play…and he maketh a rutheful noyse and gustful when one proffereth to fyghte with another.'

Bartholomew Glanvil 1389

The acceptance of the cat as a pet came about largely as a result of the rise of cat showing in the nineteenth century. Becoming a pet has brought further changes in the cat (see Chapter 3), accentuating the desire of some people for an animal showing greater dependency. Related to this is the impulse to control a cat's predatory behaviour. To this end, control laws have recently been passed in parts of Australia, and have been proposed for some areas in the US. This fear of the cat 'red in tooth and claw' does not, however, tie in with the cat's catching ability. Our current realisation is that a cat playing with live prey is not a 'wicked torturer' but a cautious animal using inhibited play techniques as part of hunting behaviour (see Chapter 5).

The first proper experimental animal behaviour work demonstrating instinctive behaviour was carried out by Galen nearly 2,000 years ago. After the debates over instinct by Descartes in the seventeenth century and Voltaire and Rousseau in the eighteenth, the early twentieth century saw a return to experimental work. Much of the behavioural research was focused on rank – establishing a hierarchy in animals. Unfortunately, the research on cats was carried out in the laboratory, not realising the changes to the animal that such an artificial environment would bring. Under these conditions, it was possible to demonstrate some rank between cats 'when the chips were down'. In a way, this was an anthropomorphic attitude (giving human characteristics to what is not human) and assumed that we knew what was important and so could disregard factors like ranges. The more recent move to field observation work on an animal whose social organisation is not particularly hierarchical is noteworthy.

I consider that for a biologist and behaviourist I have been extremely lucky, for I began my television career presenting with Johnny Morris on the *Animal Magic* programme for the BBC. In that incredibly long-running and popular series, Johnny was

most famous for putting voices onto film of animals, and was consequently criticised by the academic establishment of the time for being far too anthropomorphic. Johnny's animals not only spoke with human voices, but they had human emotions: his seals loved catching fish, his camels spoke disdainfully and his cats purred with pleasure. This may have been out of step with the then current behavioural and ethnological thought, but Johnny had a powerful predecessor in this in Charles Darwin.

Although most people now agree with Darwin's ideas in his book *Origin of Species* (1859), he also caused a great leap forward in the study of behaviour with his work *The Expression of the Emotions in Man and Animals* (1872). Darwin noted that while a quarrelling cat might be crouching with flattened ears, a cat showing affection to its owner may stand tall with arched back and tail held high, and that such postures are the antithesis of each other, corresponding with opposite emotions. During this century, using terms like

'affection' came to be seen as dangerous, for it put *our* understanding and feelings of emotion into animals that could not talk, and therefore could not be asked what they were feeling. Indeed it became 'correct' to regard animals as automatons mindlessly following instinct.

Johnny Morris' flagrant disregard of such a sterile approach was fraught with dangers in interpretation, but what never seemed to occur to the anti-anthropomorphists was that so was their approach. This disparity caused me to realise that the 'antis' were in danger of 'throwing out the baby with the bath water'. Without making the anthropomorphic error of adding in human values that are not there, we do need to be just as cautious about throwing away points of genuine similarity between us and cats (see Chapter 10). When we change our perspective we change our interpretation, and the current move away from 'pet' status and towards 'companion animal', which certainly carries risks of blatant anthropomorphism, also benefits from those gains of insight.

1 Why is a Cat...?
Essentially Feline

In its behaviour, the cat is the most enigmatic of animals – independent yet playful, lordly yet graceful, the keenest of hunters yet prepared to sleep for hours. Its dramatic sexual life led to its identification in ancient Egyptian times as the female fertility deity Bastet, and its sensual nature is recognised today by the way in which we still identify feminine with feline. A darker view of this association led to that awful period for the cat – the witchcraft era. During those changing times in the fortunes of the cat, its essential behavioural character did not alter, merely our perceptions of it.

Today, different people still perceive the cat very differently. A cat-lover will talk of its self-reliant independence, while a cat-hater will regard it as a stand-offish animal. If an anti-cat person is also a dog-lover, they might refer to that animal as loyal, while if a cat-lover is anti-dog, they might say that a dog is sycophantic.

Our ability to interpret any animal's behaviour is clearly dependent on how we view that species – and individual views vary. With an animal like the dog, unravelling this may be relatively easy, but the cat has evoked such a range of emotions and attitudes that an objective interpretation of behaviour is hard to find.

It is remarkable that we share our fireside with those two carnivores, the cat and the dog. Two herbivores – say, a sheep and a rabbit – might be considered safer. After all, if without a track record of pet keeping we were to consider bringing in wild carnivores to live with our children, concern would certainly be expressed! Yet our relationship with cats has become so intimate that the term 'pet' is now well on the way to being ousted by 'cat companion'.

Carnivores are clever and curious, and both cat and dog, to use Shakespeare's word for the cat, are 'necessary'. The dog, deriving from the wolf, is a pack animal and moves about with us, its human companions, as a member of a pack. In

The epitome of the highly trained dog: the decoy dog and handler. The dog gains its confidence from its 'pack' leader, its owner

contrast the cat, deriving from wild cats, stays around the home, held by its territory (see Chapter 4). This simple distinction pervades every aspect of both animals' lives, right down to how they interact with us. We know that a cat is different from a dog, but why?

Why should a cat have an unassailable air of independence and yet a dog do as it is told? The strategy of either group or solitary hunting is dictated by the landscape (see Chapter 2). In the dog family, packs are more successful hunters in open landscapes than solitary dogs, and the mother and litter are therefore more supported. However, in enclosed terrain, solitary hunting cats would be hindered by others and there would be more risk of disturbing the prey. Mother cats do not therefore have the support of a pack and as a result have to be more self-reliant.

Cats are not as hierarchical in their social organisation as dogs. The downside of a lot of carnivores living together is that a ranking order

Ever alert, the independent cat checks out its territory alone, and gains its confidence from the integrity of the territory

needs to be established before hunting, and the determination of this means that there is greater potential for injury. Dogs avoid this by the use of an extensive range of submissive postures, such as avoiding eye contact, wagging their tails and simulating food-begging. For the lone hunting cat most of this behaviour is irrelevant, as that situation does not arise. Consequently, and contrary to popular opinion, which historically refers to the dog as a 'simple, bluff, honest fellow' and the cat as a 'duplicitous and Machiavellian' animal, it is now the cat that has been dubbed by behaviourists 'the honest communicator'. An individual dog is a cog in a machine, while the cat is the entire machine. As a hunter the cat is supreme, but it must avoid injury at all costs, for, unlike an individual dog, it cannot afford to have an 'off-day'.

How your cat works

The body structure of the cat is intimately related to its behaviour as a solitary, semi-arboreal hunter with a partially nocturnal lifestyle. Each part of the body shows adaptations which together make the cat the hunter supreme, allowing it to stalk, kill and eat its prey with maximum efficiency. In addition, the cat is fiercely territorial, and its body is well adapted to both leaving and interpreting scent messages around its home range (see Chapter 4).

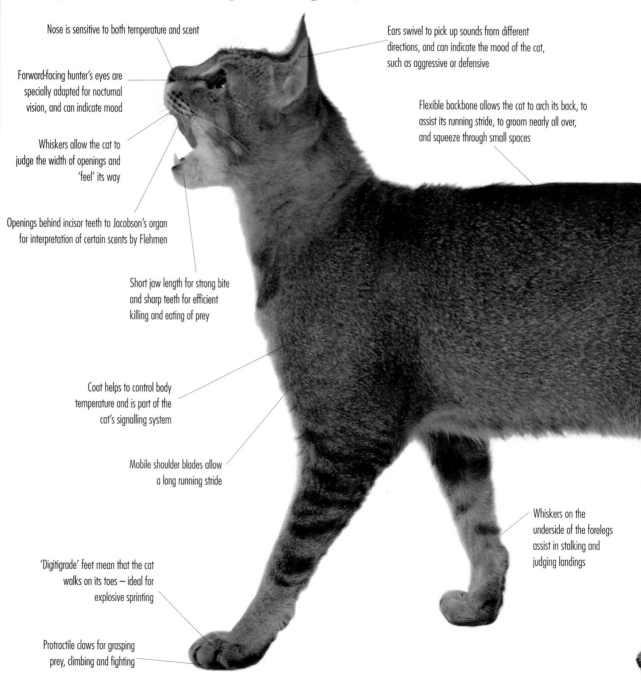

Nose is sensitive to both temperature and scent

Ears swivel to pick up sounds from different directions, and can indicate the mood of the cat, such as aggressive or defensive

Forward-facing hunter's eyes are specially adapted for nocturnal vision, and can indicate mood

Flexible backbone allows the cat to arch its back, to assist its running stride, to groom nearly all over, and squeeze through small spaces

Whiskers allow the cat to judge the width of openings and 'feel' its way

Openings behind incisor teeth to Jacobson's organ for interpretation of certain scents by Flehmen

Short jaw length for strong bite and sharp teeth for efficient killing and eating of prey

Coat helps to control body temperature and is part of the cat's signalling system

Mobile shoulder blades allow a long running stride

Whiskers on the underside of the forelegs assist in stalking and judging landings

'Digitigrade' feet mean that the cat walks on its toes — ideal for explosive sprinting

Protractile claws for grasping prey, climbing and fighting

Tail aids balance when climbing or turning suddenly, and is part of the cat's signalling system

Strong muscles in the hindquarters and back provide power for climbing and jumping. The position of the back in the cat's stance signals its intention

Scent glands below the tail, along the top of the body and on the lips and chin for leaving scent messages

Pads on the paws cushion movement and act as shock-absorbers when landing

CAT SHAPES

Compared to dogs, domestic cats are all pretty much the same in size and shape. The average adult height is around 30cm (12in) at the shoulder, with a body length of 45cm (18in) and a tail about 30cm (12in) long. There are two basic variations on body shape: the heavily built, 'cobby' type, and the lithe, lightly built type.

'Cobby' type

Lithe shape

Breeders have sought to increase the fullness of the cobby shape in the British Blue, whilst accentuating the slimness of the modern Siamese. This is reflected in their skull shapes. If not neutered the male's build becomes larger than the female, with the male's heavy jowls being most noticeable in cobby cats like the British Blue.

Movement

Walking

One of the instantly recognisable 'feline' characteristics of cats is their graceful and sinuous movements. On the aptly named 'cat-walk', elegant models parade back and forth with the same lithe beauty of movement seen in the cat. For the cat this characteristic is a survival mechanism, for as a lone hunter it must have the

PAW PROTECTION

A cat's paws have rounded pads below the toe bones and a large central one below the metacarpals and metatarsals, which cushion movement and are the prime shock-absorbers when a cat lands. For further protection, the skin of the pads has an outer layer – an epidermis – that is some seventy times thicker than elsewhere on the body. The pads of the paws are kept soft and supple by watery eccrine sweat glands. When washed by the cat, the curious pad pulling and sucking alarms some owners, but it is perfectly nor-mal. Cats are very choosy about surfaces that they are happy to walk on: sharp gravels and rough mats are usually skirted around.

advantage of flexible movement in pursuit of its prey. To facilitate this, the cat walks on its toes in a 'digitigrade' manner, which both increases the limb length and reduces contact with the ground – a necessary feature for a sprinter. (Our own large feet are required to stabilise our erect frame.) Hoofed animals like antelope and deer reduce this still further for speed, but the cat cannot sacrifice paw manipulation (see page 14).

Running

The cat is an explosive sprinter – it does not go in for long chases. By having its shoulder-blades aligned on the side, with only a vestigial collarbone, the shoulders are free to move, which increases the running stride. When walking at a slow pace there is a naturally retarding action when the forelimbs are placed down, but when running fast nearly all that effect is lost, for the cat extends the forelimbs and arcs down and back before contact is made with the ground. The flexible arching of the cat's spine also allows it to extend its stride further by several inches.

Cats alternate opposite paws while walking, but when running and galloping the movement of the

Left: Sixty per cent of the cat's weight is carried by the forelimbs during walking, and consequently they provide proportionately more support, while the hindlimbs primarily provide propulsion

Far left: When a cat jumps, whether onto a table, branch or prey, it first takes all its weight onto its hindlimbs, and it is the powerful extension of these that propels the cat's leap

legs changes so that the legs are used together in turn on each side of the body. When in full gallop the cat is airborne for most of the extended stride, without any paw touching the ground.

Jumping

The strong muscles of the hindquarters and back give the cat its tremendous power to jump up, down, and over a gap or obstacle, provided it has a firm surface from which to push off. A cat can leap several times its own length either vertically or horizontally.

When a cat is at full gallop, its forepaws, after landing, between airborne bounds, are overtaken by the hindpaws

The cat will size the jump carefully for some time and test the firmness of the take-off with its hind feet, before making a perfectly judged leap. This patient assessment is crucial when the landing place is small or narrow, for example a shelf, window-sill or tree branch, or the gap to be cleared is wide. It is also important when the cat pounces on its prey: here, the judgement is of where the moving prey will be when the cat lands.

Teeth, Tail and Claws

Teeth and Claws

The reason that paw manipulation is important in the cat is that it has sacrificed skull length in favour of a more powerful bite. However, one result of this gain is a flatter face and loss of visual information from in front of the mouth. A cat can see an object clearly at a reasonable distance, but it disappears just in front of its mouth. As we too have flat faces, we share this problem. We also share the answer – flexible front feet (or hands).

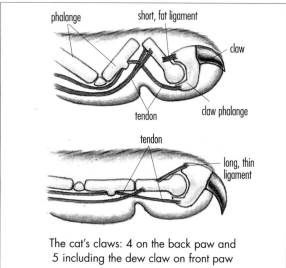

phalange

short, fat ligament

claw

tendon

claw phalange

tendon

long, thin ligament

The cat's claws: 4 on the back paw and 5 including the dew claw on front paw

Cats' claws allow them to move with ease and confidence from one tree branch to another

structure and behaviour are intimately linked.

The claws are also useful crampons for the semi-arboreal cat. For climbing, as well as for prey-catching and fighting, the effectiveness of the claws is increased if they are sharp, and this is the main reason why cats claw trees. As they drag them through the bark, curved slivers of keratin

Cats can grip with their front paws with more than just flexibility: they have protractile claws like blades that simultaneously fire from each paw! (To call them 'retractile' is to misunderstand the animal's behaviour.) In a normal relaxed state the claw is sheathed, but when the paw is extended ready to strike (in a similar move to our opening our hands wide) the curved claws project. Consequently, while a wolf or dog's first contact with its prey is with its teeth, a cat's is with its protracted claws on extended forelimbs. As ever,

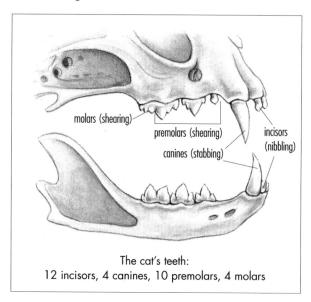

molars (shearing)

premolars (shearing)

canines (stabbing)

incisors (nibbling)

The cat's teeth:
12 incisors, 4 canines, 10 premolars, 4 molars

flake from the sides of the claws, bringing them to a point again. Keeping the claws sheathed most of the time protects them from being worn down.

Once prey has been caught and killed, the cat's eating equipment is also that of a dedicated hunter. When Linnaeus referred to the long teeth of the cat as 'canines' it was an oversight, for they are more dramatic in the cat family than in the dog. Due to its hinging, the cat's jaw has virtually no lateral movement and this makes the gripping bite effective. The number and formation of the teeth also assist in making possible a shearing action.

Cats have a reduced number of molars and the premolars are aligned like serrated scissors to cut through flesh. They have no facility to grind food. It is because of this that cats adopt a strange, gulping manner when eating grass and leaves.

In mid-chase a cat can corner with feet scarcely touching the ground as the tail counterbalances movement

The Tail

The cat's tail is an incredibly versatile organ. It is of particular value in the balance of tree-climbing cats, but also acts as a gyroscopic counterweight when cats need to corner suddenly after prey. In addition, it is also a signalling system and can be fearfully fluffed, indecisively twitched or aggressively wagged. When an intact tom is spraying the quivering tail is held high, keeping it well clear! Below the tail, on either side of the anus are scent glands activated by hormones that are used to leave scent messages (see Chapter 4).

AIR RIGHTING AND BALANCE

As tree-climbing animals, cats have a remarkable and vital ability to land on their feet by rotating their bodies in mid-air. This is a reflex action, appearing in the kitten in the third week of life as mobility increases. As it falls through the air, the cat first rotates its head and the front half of its body, until its head has achieved the correct orientation. The cat then rotates the back half of its body, allowing it to land safely on its feet.

The cat is able to carry out this manoeuvre due to its finely attuned sense of balance. It is this same exquisite sense of 'uprightness' and movement that feeds the cat information on its posture during the dramatic changes of position which take place in the course of a bird capture (see Chapter 5) or a fight (see Chapter 8). This awareness is achieved by both vision, and canals in the inner ear (see page 18).

The cat's tail curls neatly round him as he balances perfectly on the fence

Cat senses

Vision

The eyes are one of the most important keys to understanding the relationship between behaviour and structure in the cat. The eyes, and especially the degree of dilation of the pupil, are particularly indicative of mood (see Chapter 9). Their strange reflective ability made cats seem magical to our ancestors, yet this is just part of the package of night-hunter specialisation. The eyes are forward facing, as befits a hunter needing the advantage of depth perception, and so the cat has good three-dimensional and distance evaluation. In contrast, herbivorous species often have all-round vision, which enables them to see a predator approaching from any direction but lacks the advantages of binocular vision.

Left: Cat in low light or fearful
Right: Cat in bright light or threatening

As with other nocturnal hunters, the cat's eye is huge relative to skull size and compared to that of daytime animals, including us. The lens itself and the cornea in front of it are large relative to the back of the eye. The lens is also set further back from the front of the eye: this is why if you look at your cat's eye in profile it can seem quite 'glassy' compared to your own. This gives the eye of the cat a wide aperture and greater light-gathering ability. At low light intensity the pupil becomes huge to allow in more light. With all these adaptations combined, the cat can see with a sixth of the amount of light that we need.

This maximising of light gathering makes the cat's eye potentially vulnerable to intense dazzle during the day, which is why the pupil closes to a slit instead of a point. This gives the cat much finer control over the gradual closing of the iris. Lions, which hunt mostly during the day, have less nocturnal adaptation and consequently their pupils close to a point, like ours.

In the retina, the cat's final visual adaptation to its nocturnal lifestyle is to sacrifice much colour

The familiar 'lit-up eye' of the nocturnal cat. The biological mirror backing a cat's eyes is made of aligned refractile rodlets crammed into ranks of special cells that are layered up to 15 deep

The shining-mirror effect of the eyes of a cat is due to a crystal mirror, the tapetum lucidum, located behind the retina. In very low light conditions, photons of light stand a greater chance of hitting a light receptor if they are reflected back from the tapetum after passing through the retina.

Circular pupils for daytime living animals controlled by circular fibres cannot fully close to zero aperture. However the slit pupils of the nocturnally adapted cat that are pulled together by crossing fibres can close completely

discrimination in favour of maximising light reception. Like us, cats have rod and cone receptors. To achieve colour interpretation, there have to be three separate colour receptors to register one visual event, but when colour is disregarded more definition can be achieved by more rods. Cats have six times the number of rods for each cone that we do. However, they do retain some degree of colour vision, but after dusk when the landscape is drained of colour (even for us with our better colour vision) the hunting cat has good night vision.

Touch

Having sensory receptors in the skin, the cat is aware of touch when its guard hairs brush against an object. It also receives specific sensory feedback from the dense network of nerve endings connected to the shaft base of the cat's whiskers. These are sensitive enough to pick up air movements. They also help the nocturnal hunter move around in woodland conditions and are just as useful in helping the cat to negotiate gaps in fencing. In particular, the muzzle whiskers are brought into play when a cat is in close contact with small prey, and when the whiskers are pulled towards it. A set of tufts of whiskers that are usually overlooked are those on the underside of the forelegs, which assist in stalking and in gauging landing from a leap.

The facial whiskers on the cat's muzzle also sweep into new positions to declare the mood and intentions of the cat, and other cats have no difficulty reading the signs.

THE COAT

Temperature control is fundamental, affecting where a cat will sit in its range and even how it sits (see Chapter 4). A cat's coat is vital to trap body heat. There can be up to 200 hairs per square millimetre of which 150 will be down hairs, 47 awn hairs and 3 guard hairs. The guard hairs and awn hairs protect the cat against the bleak elements. While guard hairs grow singly, both the awn and down hairs grow in clusters that emerge from single hair pores. When a cat moults to produce the changes of coat density to match the season, the new hair grows in the same follicle shaft and forces out the old hair.

The base of the cat's whiskers go three times deeper into the skin for firm attachment than do the longest cat hairs. They are bedded into individual fibrous capsules attached to their own large arrector muscles which let the cat sweep its whiskers forward to investigate prey or another cat, or pull them back out of the way. The base of the whisker has four types of nerve receptors so when the whisker is deflected the degree of pressure, the direction, speed and duration can all be felt accurately by the cat. They let the cat not only interpret close prey movement and shape, but even the direction of its coat

Cat senses

Hearing

If your cat is standing in front of you, apparently oblivious to what is behind it, look at its ears and you will probably find that they are partly pivoted towards you. At a sound elsewhere, the ears will instantly position more definitely.

The cat has over twenty muscles that work the pinna, or cone, of the exterior part of the ear. The pinna acts like a pivoting ear trumpet, swivelling to pick up the slightest rustle. The cat can pinpoint the sources of sounds much more accurately when it is stationary than when it is on the move, so will often stop and listen carefully. The cat uses the same muscles to move its ears in a variety of signalling positions, to convey its moods and intentions (see Chapters 8 and 9).

Sound is funnelled down from the pinna to the eardrum and vibrates against a group of small bones in the middle ear. These pass on and amplify the vibrations to another drum at the base of the fluid-filled cochlea in the inner ear. Vibrations are detected by hair cells in the cochleal walls. The cat has an observable operating upper limit of 60kHz.

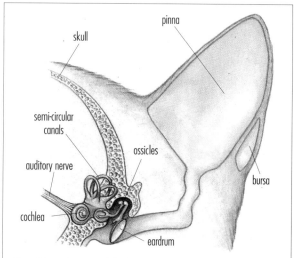

Most of the working ear is lodged in the skull. The small ear bones, or ossicles (known as the hammer, anvil and stirrup) exert leverage against each other in response to eardrum vibrations. Their airspace extends into the auditory bulla chambers

Part of the great flexibility of the cat's ear, vital for conveying mood as well as detecting sound, is provided by the bursa, the almost unnoticed pocket at the base of the outer edge of the pinna

This is significantly higher than the operating upper limit of both dogs and people, encompassing the high-pitched sounds of rodent squeaks, which are in the 25kHz range. The large bulbs positioned at the back of a cat's head at the base of the skull are the auditory bulla, and it is thought that these are particularly sensitive to the sounds made by rodents.

The cat's acute sense of balance and movement is achieved through its vestibular system – the group of three semi-circular canals in the inner ear, embedded in the skull. These are filled with fluid, which tends to stay in position through inertia, despite the twists and turns of the cat. Projecting from the walls of the canals are sensory hairs that detect the surge of relative movement between the cat and the fluid. The canals are set at different angles, so movement in any direction is picked up. When a cat is falling (see page 15), it needs an awareness of gravity when it is the right way up, and this is provided in part by calcium particles in the fluid in the canals, which land on the haircells.

Scent and Taste

The cat's nose is filled by layers of bone that are covered with a mucous membrane, which is twice the area of ours. This indicates the relative importance of scent to the cat, for whom it is an essential tool in identifying strangers, marking group members (which can come to include us) and interpreting scent messages left around the home range by other cats (see Chapter 4). It is

The ability to detect scent is greater when air temperature is lower than ground temperature, as occurs in the evening

also important in determining the cat's response to food before tasting.

In tigers as well as tabbies, the middle of the tongue is covered with backward-pointing spines that act as a rasp to break off and assist in gripping meat. The taste receptors are positioned on the tip, back and sides of the tongue only. Most mammals can interpret the range of sweet, sour, salty and bitter tastes, but the cat – as a pure meat eater – has hardly any sweet receptors. A cat's digestion can be upset by sweet foods – if it will eat them at all.

When a cat licks us we become aware of their tongue's rasping spines. A close-up scanning electron micrograph show the huge gain in surface area they give, helping the tongue to pick up fluids

THE EXTRA SENSE (FLEHMEN)

Cats have an extra organ that we do not have which helps them to interpret scent. They can 'taste-scent' using the Jacobson (or vomeronasal) organ, which is named after the Danish doctor who discovered it almost 200 years ago. This lies below the floor of the nasal cavity and opens to the mouth just behind the first incisor teeth. As it consists of a pair of blind-ended sacs, air cannot enter into them to be assessed without the assistance of a little breathing control.

To do this, the cat adopts a curious, grimacing pose with its mouth partly open – the 'Flehmen response' – which closes its normal breathing route and instead draws air through the ducts behind the incisor teeth, allowing the scented air to be checked. Sometimes the tongue flickers to aid wafting of the scent. The tiger's Flehmen response to urine is most dramatic, as it pulls back its lip and exposes its huge teeth, but that of the domestic cat is more subtle, so much so that most owners have never noticed it.

A wide range of studies across many species suggest that the main use of Jacobson's organ is in urine analysis, and it is usually employed by males to test the sexual condition of females. Because of the territorially dispersed nature of cats, it is important in the timing of mating for males to be able to evaluate a queen's sexual status accurately. Although other males from further afield will eventually gather information on the queen, the resident males will have the advantage, being on the spot to detect early changes in pro-oestrus, ahead of non-group males. When a queen is already in oestrus, the rolling and treading behaviour that toms evoke in her can also be stimulated by the urine spray mark of a tom.

2 From Tabby...
Build for Behaviour

Cats are carnivores, and of all the mammalian families the cats epitomise the hunter supreme. Their structure and function are so bonded together that cats are instantly recognisable as 'cat', whether they be tiger, cheetah, Margay or Pampas cat. Closely tied to their appearance is their behaviour, which remains remarkably constant throughout the family Felidae.

All domestic cats descend from the original tabby that emerged at the beginnings of domestication from its wild ancestors. Despite their appearance, even Siamese or pure white true Turkish Van cats remain genetically tabbies. Tigers and tabbies share many feline family features, but in one regard they are a

The tiger has a massive and powerful front end to pull down large prey and a strong neck to twist it around and expose the throat. Strong jaws with huge teeth grip the throat of large prey in an asphyxiation clamp bite while its huge front paws allow it to secure a good grasp on its prey. But an imbalance in the size of its front and back ends mean that it has a reduced sprinting and tree-climbing ability

...to Tiger

genuine duo, for they are the only fully striped cats. When they hunt, both will stalk their prey. With every muscle tensed, they creep slowly along, their bodies low to the ground, their eyes looking dead ahead, unblinking.

The tiger's prey may be a Chital deer and the tabby's a young mouse, but for both cats the hunt is just as much in earnest. In both cases, if the prey spots them the cat will lose not just its immediate prey but also others, as the alarm would be given. In consequence, both advance by short moves, freezing if there is any risk of spooking their potential prey. When they judge the time is right, both will suddenly rush and leap upon these prey. Both despatch their prey by bites to the back of the neck.

The cat's balanced build is ideal for sprinting and climbing – its strong back end provides power for climbing. Its paws are in proportion as prey is relatively small and its sharp teeth are ideal for killing small prey by a bite to the back of the neck

Physical Specialities

Where physical specialisations do occur in the cat family they have accompanying behaviour and habit differences. Most cats are good climbers, and the balanced build of the leopard allows it to avoid ground scavengers by hauling its dead prey up into trees. The build of the tiger restricts its ability to climb, and its disproportionately powerful front end allows it to pull down large prey following a jungle stalk. The cheetah, which again has a limited climbing ability, has a much longer tail than most cats, that it uses gyroscopically for balance in the chase on the open plains; canine teeth size has been sacrificed in exchange for larger airways to maximise oxygen intake when sprinting. These gains have been at high cost, for hyenas regularly take cheetah kills. Yet cheetah cubs are able to climb as youngsters. Small cats catching small prey do not generally have this problem of losing kills, as most prey can be eaten in one meal.

The ultimate specialist in tree climbing is the Margay, which lives in the rain forest of South America. The protractile claws that cats use as built-in crampons allow most of them to climb up trees fairly easily, but they are less able to clamber down with the same grace as the claws are one-way hooks. The unique Margay is able to walk down a trunk as easily as it climbs up due to the flexible ankles in its hind legs, which allow its back feet to be reversed.

Top left: The margay – the ultimate tree climbing specialist

Left: The perfectly balanced build of the leopard – a larger version of the domestic cat. Its broad shoulder-blades, with solid muscle attachments aid climbing. Its balanced frame allows the leopard to climb easily – and it can haul its dead prey up with it, away from ground scavengers.

Above: The slender overall shape of the cheetah with its long limbs allows it to achieve speeds of up to 112kmph (70mph). The canine teeth are greatly reduced allowing the airways to be enlarged for maximum oxygen intake and its very long tail, which has 28 vertebrae (far more than most cats) rotates rapidly for balance when turning suddenly after prey. However this tremendous speed does not come without a price and its slender shoulder-blades, needed for speed, mean reduced climbing ability

Cats Large and Small

From most documented evidence, our house cats owe their origins to domestication in ancient Egypt, and even today the cats of the warm Mediterranean and those (following historic translocation) of hot south-east Asia are of the same slim build. In northern climes, domestic cats are heavier. Pumas of the American tropical rain forest are significantly lighter in build than those of colder southern South America, and North America. The same climate distinction occurred across the range of tiger sub-species, from the massive Siberian to the smallest (and now extinct) equatorial Balinese and Javan tigers. The adjacent population of Sumatran tigers are now the smallest. Living on islands has accentuated this tendency.

Many cat taxonomists recognise thirty-seven or thirty-eight species in Felidae, of which thirty-seven are wild. However, this is an ever-changing minefield! One thing is certain – all cats are efficient hunters. As could be expected from that number of species, some cats overlap each other's distribution – so are they in direct competition with other members of the same family?

The main ancestor of the domestic cat, the African wildcat in its spotted form

The even-coloured (agouti) Jungle or Marsh cat may be a part of the domestic cat's ancestry

DOMESTIC ANCESTORS

From radiotracking of Forest or European wildcats in France we now know that they have a similar range pattern to the domestic cat, and this also appears to be true of African wildcats in Saudi Arabia. Nonetheless, we know remarkably little about this cat, the main ancestor of the domestic cat. The African wildcat seems to be a paler southern variant of the Forest wildcat, and there are spotted and striped forms. It is thought that those nearer human habitation are more tractable, but extensive hybridis- ation has occurred for centuries between these cats and domestic cats, which may be responsible for this behaviour.

The Jungle or Marsh cat has also been implicated in contributing some genes during the early period of domestication. A gentle animal, it was kept captive at cat temples by the ancient Egyptians, and its mark- ings are similar to those of the agouti Abyssinian. These cats seem tolerant of people, and I have found them living happily near villages in South Nepal.

Remarkably, many factors reduce such competition, one of the key ones being the size of the cat. The big cats like tigers are clearly distinguishable from small cats like the domestic cat by size alone (although taxonomically the

YOUNG CATS

The young of most cats are cared for exclusively by their mothers and are kept discreetly hidden. Camouflaging colouring helps here. All wild cats have visible patterning when born, including those where adult habitat adaptation has lost pattern in favour of becoming sandy, as in the lion and puma. Cubs that need to follow their mothers to kills, such as lions, tend to be born in a more advanced physical state.

puma is grouped with the small cats!). There can also be more ambiguous groups of mid-sized cats. Small cats catch only small prey, while big cats catch some smaller prey but also large prey that the smaller cats ignore. (There is more on the killing of prey in Chapter 5.)

The Smallest Cat

The smallest of the small-cat species is the Rusty-spotted cat which weighs only around 1kg (2.2lb), which is half the weight of a small domestic cat. It lives in parts of India and Sri Lanka. Its size suggests that it exploits its woodland habitat by a semi-arboreal life, but in fact we know little of its behaviour. Not all small wild cats are forest-dwellers, the most notable exceptions being the Pallas and Sand cats. Both of these are physically adapted to their specialised habitats.

The grumpy-looking Pallas cat lives in the mountains and high plateaux of central Asia. Its short legs have given it a remarkable facility to move around on rock faces, while its long thick fur protects it from the cold. The Sand cat is a small cat that looks even smaller than it is because of its large ears and broad skull. In its way it is as specialised as the cheetah. It survives even the Sahara Desert by having long tufts of hair between its footpads and being strictly nocturnal. The large ears help it catch its prey; insect prey are important to it as a source of water. This is the one cat that is a truly competent burrower, having been found to have excavated a burrow nine times its own body length.

Sadly, we know very little about the behaviour of many small wild cats, but as we have only been studying the behaviour of feral cats in any detail for around twenty years, this should not be too surprising. For example, despite its name the South American Pampas cat lives not only in grassland but also in cloud forest. Although biologists know little about this animal, during the late 1970s 26,000 Pampas cat skins were exported every year until it became a protected species. Clearly there were some people who knew exactly where it lived. We do know, however, that it is one of the few cats spoken of as 'untameable', a characteristic also attributed to the Scottish wildcats.

The small Sand cat's main prey are desert rodents and reptiles

The Pampas cat is said to hunt ground-nesting birds and guinea pigs, but we know little of its behaviour in the wild

Habitat Preference

The colouration of lions blends them into the arid landscape, and that plus co-operating as a pride make hunting large prey a successful strategy – despite that potential prey know they are safe as the pride rest after a meal

Solitary or Group Cats?

The normal pattern of hunting behaviour for the cat family is solitary hunting by stealth. The success of this technique depends on cover, and so the usual landscape for the cat family is one of forest or dense grassland.

Coat patterning is a reflection of habitat. The strongly dark-spotted cats, like the Margay (see page 23) or Clouded leopard, blend into woodland. Although at ease on the ground, these are the true arboreal cats of the New World and Old World respectively. Both have long balancing tails, the Clouded leopard exceptionally so, countering its heavier build. The Clouded leopard has the profile of a large cat, with proportionally the largest canine teeth of all cats – the opposite of the cheetah. This is consistent with having a long solid neck and relatively short front limbs. In trees, it obviously cannot trap birds by the normal method of leaping and pulling them down to the ground (see Chapter 5). I would suggest that its bigger teeth indicate that it uses its mouth proportionally more to compensate for its shorter limbs. It has a small-cat size body, and purrs like a small cat. For both of these extreme woodland cats, their habitat-based lifestyle would make group living impracticable.

The tiger's imbalance of build not only enables it to bring down large prey by itself, but is a dramatic verification of its stealth technique, for it does not have the physique of a runner. In contrast, George

URBAN 'LIONS'

Feral cats show a flexibility of social grouping that relates to food availability in the landscape (see Chapter 4). There are some parallels between prides of lion and the feral cat groups focused around an urban waste heap – mainly larger defending male ranges and a predominance of direct female lineage. In both cases, daughters are likely to stay around and the females of the group will be mainly sisters, cousins, aunts, grandmothers and so on.

Schaller, the premier ecologist of big cats, found that in a particular habitat a single lion caught prey on 29 per cent of hunts, but when part of even a small group, the rate increased to a far more successful 52 per cent. He also found that in open terrain with little cover, a solitary lion had half the success rate it had when hunting in long grass.

The normal lion pride size is around fifteen or so, in which the females do the hunting while the males defend the pride's territory. Group hunting enables the females to tackle large prey like wildebeest and zebra regularly. At one time the big cats were distributed more widely than they are today, and the areas they have survived in most successfully are those to which they are best suited. The plains of Africa are the heartland of the lion, but a small pocket remains in the Gir Forest in Gujarat, India. This is an area of dry open woodland, but with far more cover than that enjoyed by the lions of east Africa, whose habitat is typified by the Serengeti Plain. Consequently, individual stalking can be more successful and so their social system is different, the males usually joining the female groups only when the latter are in oestrus. The normal prey is also smaller than in Africa, usually consisting of Chital deer.

Open Ground or Cover?

An open plain landscape does not just affect the survival 'strategy' of predators but also that of the herbivorous prey species. The open grassland plains of Africa support huge herds of animals, at high densities. The numbers provide a buffering protection for individuals and result in a sacrifice of the frail to large cat predators. The herds also increase the probability of a kill for the cats and decrease the need for stealth, as used by the tiger in its more enclosed habitat. The lion takes advantage of the large numbers of prey by co-operative pincer moves that create confusion in the herd.

Cheetahs have turned primarily to speed in order to outrun antelope and gazelle on the open plains. Their celebrated speed is necessary, for some antelope can shift at incredible speeds – only a few kilometres per hour less than the cheetah. The social system of the cheetah has also been found to be dictated by the landscape. Although a significant part of the male cheetah population hunt alone, over 60 per cent hunt in groups of two, three or more. This different tactic does not improve their success rate, but it does increase the prey size that they can reliably kill – from, typically, a Thomson's gazelle to a wildebeest that is four times the weight. These groups of males also find it easier to take over and hold territories than do single males.

Cats that use cover are more typical of the cat family. The build of the tiger enables it to deal with large prey unaided, and the dense cover it favours makes hunting co-operation impractical anyway. In addition, the tiger's prey does not form large herds in its woodland or marshland habitat.

Flexible or Fixed Territory?

Due to the seasonality of available grazing, the dry African plains are witness to massive annual migrations of wildebeest, which in turn affect cheetah and lion mobility. In contrast, tigers are more fixed in their territory, scent-marking it in the same way as domestic and many other cats (see Chapter 4). In Chitwan, south Nepal, radiotracking work has established that, like the domestic cat, the male tiger holds a much larger home range than the female. As food in their habitat of jungle and marsh is spread out, the female home ranges do not overlap significantly with each other, but a number of female ranges come within the range of one male. Where conditions are optimal for tigers, with abundant prey as at Chitwan, their density is around six times higher than that of Siberian tigers. These general patterns seem to be typical of most of the cat family.

The leopard, with its general big-cat build, is less habitat-restricted than most big cats, and consequently has the most widespread distribution. As might be anticipated from this, it has the widest range of home-range sizes and densities. In the Americas, the jaguar has been radiotracked (although not as extensively as the tiger), but nonetheless the pattern of territoriality, with males holding ranges larger than females and a number of females being within a male range, seems to be the same. Similar radiotracking of the most widespread cat of the Americas, the puma, found that the degree of overlap of female ranges varied considerably. Summer ranges were also larger than winter ones.

Territory and ranges for domestic cats are discussed in detail in Chapter 4.

The Clouded leopard has a disproportionately long and heavy tail to counterbalance its movements when hunting amidst tree branches

3 Breed Behaviour

Generally, there seem to be fewer specific behaviour patterns in cat breeds than in individual dog breeds. This is partly due to dog breeding and showing having a longer history than the cat equivalent, but also to the closer similarity of many cat breeds than the wide range of shapes and sizes seen in the dog arena. Behavioural differences that do exist between cat breeds are often due to their differences in build.

Build and Behaviour

The biggest differences in breed temperament and behaviour are between the main historic breed groupings, where the fuller-figured and heavy-coated Persians are significantly more lethargic than the extrovert short-haired cats of south-east Asia, such as the Siamese. Generally, the British, European and American short-haired cats fall between these types. However, the traditional form of the Siamese is nothing like as extreme in its liveliness as the modern, thinner form, which is highly active and can whiz around as if it is a firecracker! Similarly, the modern exaggeration of the Persian form, with its flatter face and fuller coat, has led to a cat that *cannot* be as active. Although Persians will sit placidly for ages, they are not great lap cats, for their abundance of fur makes them uncomfortable on our warm laps and they overheat.

Genetic Inheritance

The main reason for this broad breed behaviour difference between the Siamese, Persian and common moggie types is that for centuries they were self-contained, with their genes separate.

The moggie is the house cat, the pet, the street cat, the farm cat. In Britain and Europe, North America, Australia and New Zealand, the cats have a common European solid build. In the Mediterranean, and south and south-east Asia, the warm climate has kept the build closer to its Egyptian origins. Due to the fear of witchcraft, the cats of Britain and Europe were not tampered with for the last few hundred years, and in consequence have an identifiable shape.

The cats of south-east Asia probably started from a few individuals brought by Arab and Indian sea traders from the Mediterranean. The predominance of the kinky tail gene throughout the region supports

The Siamese is often the choice for those wanting lively and demonstrative cats, but wool sucking and obsessive overgrooming, producing hair loss, are also associated behaviours

Above: In the USA, Burmese are kept as traditional brown, but in the UK a range of colours are bred. Despite that, their distinctive behaviour pattern remains, similar to Siamese, but not as extreme
Left: The placid temperament of the Persian today is in stark contrast to those in the early days of cat showing. A century ago Harrison Weir found it to be 'unreliable'
Right: A classic Turkish Van cat, bred from the cats from Van, retains the gentle playfulness and build of its ancestors

this idea of a few founder cats. The Persian is of more recent origin, in that ancient Turkish Angora and Angora-like cats across the eastern end of the Mediterranean to the Caspian Sea were fused together in eighteenth-century France and then nineteenth-century Britain to produce the long-haired amalgam. In consequence, each of these major groups is genetically separate from the rest. They are distinct in appearance and build and have some identifiable behavioural differences.

A survey of American show judges and another of British Veterinarians identified distinctive breed behaviours and confirmed popular opinion. Generally, Siamese, Burmese and Oriental Shorthairs are the most active, outgoing, excitable, demanding, vocal and destructive. The Siamese's attention seeking is particularly noticeable due to their most vocal nature and their way of vocalising, which is often described as 'talking'. Persians on the other hand have been found to be lethargic, reserved, inactive, and not demanding of affection in the manner of Siamese. They are happier being petted than held on a lap.

Build differences even affect playfulness, with the Siamese being the most playful and Persians the least. Persians also do not score well on being affectionate. One difference which emerged from the two surveys was that the vets found that Siamese did not like being handled, while the judges said the cats loved it.

Himalayans (Colourpoint Longhairs), have an intermediate behaviour pattern between Persians and Siamese, which is more consistent with their ancestry than their primary build. They are not as outgoing as the Siamese or as reserved as the Persian, but their full coats do make them just as unhappy about being a lap cat as is the Persian. However, behaviour differences are not due just to long hair versus short hair, for there is a distinct difference between the British/European/American Shorthair, and the Siamese/Burmese group. Similarly, the difference in temperament and activity patterns between the robust Northern Longhairs, such as Maine Coons and Norwegian forest cats, the lithe, playful real Turkish Van cats and Angoras, and the more plodding Persians, is considerable and relates to build.

The Abyssinian and Somali have been found to be active, demanding of attention, yet fearful of strangers. They are often described as quiet, yet can be very vocal. The Russian Blue is possibly the shyest of breeds. It is quiet and prefers to avoid cats or people that are strangers.

Physiology and Behaviour

There are physiological differences between Persians and Siamese which mean that the onset of puberty is later for Persians and they normally have a lower mean number of kittens per litter. Kittens of the ancient Turkish Angora cat open their eyes earlier than the modern Persian. Siamese have been noted to show more paternal behaviour than most cats, in that they will lie with and groom the kittens. The mix of placid Persian and sizzling Siamese in the hybrid Himalayan has produced not only a blended behaviour but a halfway house in the litter size between the small Persian and the prolific Siamese.

Inbred Temperaments

Researcher Bonnie Beaver has suggested that the Abyssinian, Russian Blue and Siamese may be experiencing temperaments exhibiting nervousness, restlessness and unreliability due to over inbreeding. She has also postulated that the smaller gene pools of rarer breeds produce more asocial cats.

Moggies versus Breeds

Moggies are the everyday house cats of Britain, the ancestors of the British Shorthair show-cat, and have their equivalents in America, Europe and Australia. Breed or pedigree cats make up only around 7 per cent of the cat population in Britain, Europe and America, yet they form the basis of about half the cat problems seen by animal behaviour practitioners. This may be because the breed cats have a narrower genetic base, and really do have more behavioural problems. However, breed owners do appear to seek help more readily than most moggie owners. In addition, the majority of problems seem to arise from keeping cats captive or at too high a density, and this occurs far more with breed cats than with moggies.

CAT CHAT

There is a range of vocalisation between breeds. The most notable are the Siamese, which are so distinctive that they are referred to as 'vocal', 'talkative', 'loud' and 'demanding', particularly when calling on heat. In contrast, Persians are much quieter, as are other 'gentle' breeds like Korats, Russian Blues, Abyssinians; British and American Shorthairs have a mid-range voice.

Early conditioning and habituation to people during kittenhood make a big difference to how cats relate to us in their adult lives, so there must be some caution when generalising about patterns of breed behaviour. Contrasting some specific breeds that have made an impact on the showbench illustrates some stereotypic breed behaviour. Each of these breeds is physically sound, and for differing reasons has a genetic vigour.

The Maine Coon

Maine Coons have recently soared in popularity. They are a no-nonsense cat, as they should be with a history of being farm cats in rugged New England. They are easily the largest of breed cats – a really fine tom matures into his full coat and build at about four years old, and can then weigh as much as 11kg (25lb)! Breeders and owners typify them as 'gentle giants', for they have an easy-going nature combined with the confidence granted by their size. They are noted for not only patrolling their territory, but also seeming to thrive on accompanying their owners.

An active working Maine Coon cat still living on a farm in Maine as have generations of its ancestors

The Singapura

At the other extreme, the Singapura has only recently been recognised as a breed, having been introduced into the USA from Singapore in the 1970s. It became known as 'the drain cat' because it was said to be so small that it could seek refuge (in the dry non-monsoon seasons!) in the island's drainpipes. It had the distinction of being in the *Guinness Book of Records* as the smallest cat breed in the world, with adult queens weighing only around 1.8kg (4lb), and the toms a couple of pounds more. It was bred initially from just four cats, and care has been taken to avoid narrowing the genetic base too much.

In appearance, Singapuras are agouti marked (or 'ticked'), with upper foreleg banding similar to non-breed Abyssinians, but of smaller size. They are gentle and calm cats which move with a quiet manner, but can be playful and have a wide-eyed openness.

Wild-Domestic Hybrid – The Bengal

The Bengal cat is a cross made during the late 1970s between a wild Asiatic Leopard cat (*Felis bengalensis*) and a domestic cat. It was recognised as a new breed in 1983. Its stunning spotted coat, in which it has inherited some of the looks of its wild parent, has led to a rapid surge of popularity and there are already over 9,000 worldwide.

It seems that virtually all species of small wild cat can hybridise with domestic cats. There have been many, but most have produced spitfires! The Bengal cat has – or seems to have – more tractable behaviour. Responsible breeders recognise that the cat will need to be monitored very carefully for wild behaviour traits. Certainly the lively, investigative nature of these cats has been a feature that has attracted breeders.

To establish the hybrid strain as a breed, only stable, friendly cats were selected and wild, timid and aggressive-natured cats were culled. The first generation hybrids were rather 'stand-offish', and not good pets. The behaviour changed with progressive hybrid crosses. Intriguingly, the early hybrid crosses often preferred to eliminate in running water, a behaviour attributed to the wild Leopard cat. Bengals still have a particular fascination with running water, which is more marked than in fully domestic cats. Turn on a tap in a household of Bengals, and it is as if a magnet has appeared.

The Bengal was first recognised as a new breed in 1983 by TICA, and its 'personality', consistently notable in eight generations, is one of pronounced investigative curiosity, restless liveliness and a dedication to hunting and chasing. The voice, not frequently used, is harsh.

Bengal cats share the same remarkable fascination with running water exhibited by their recent wild antecedents. Nonetheless breeders hope that the new cat is primarily domestic

Behaviour Restrictions with 'Designer' Cats

In recent years there has been an unfortunate move to accept novelty or, as they are sometimes called, 'designer' cats. By choosing to select some of these animals as breeds – to the disadvantage of the animals – the health and welfare of the cats have been put in jeopardy, and the physical changes have not been without behavioural cost. These changes have occurred mainly in two ways.

Selecting Extremes

The first is by gradual selection towards exaggerated extremes, in large part caused by awarding points for extremes at the showbench. This has produced the modern Persian's fuller, flatter face, build and heavy coat. It has also formed the thin, snippy, contemporary Siamese.

The most exaggerated form of the flat-faced Persian is the Peke-faced Persian, although the 'ultra-types' in Britain are not far behind. In veterinary terms they have maxillo-facial compression, with a predisposition towards respiratory, pharyngeal and eye diseases. The contorted passageways and change of build

The thinner contemporary
Siamese cat

with the full coat make exertion less attractive to the cat. Persians generally are recognised as being more lethargic; while they can and do run and play on occasion, their activity profile is more sedentary. However, this activity reduction cannot be attributed fully to a long coat, for Northern longhairs such as the Maine Coon are active, with much patrolling activity over their territories.

The change from the traditional standard-shaped Siamese to the thinner form has produced more restlessly active cats. These modern Siamese are light in build, but they have also undergone a change in the shape of the skull, which has been extended and narrowed. However, there have also been suggestions that the behaviour change may not be due entirely to build differences.

Radical Change

The second change in recent breeding practice has been a greater acceptance of radical appearance changes arising from mutations. A number of these are controversial, not least because of their behaviour implications.

The Scottish Fold, with its ear deformity, must be outcrossed to avoid horrific cartilage and bone deformations. The Fold's ears are permanently flattened in a defensive look. The American Curl from California similarly has ears permanently distorted back which may seem to another cat to be signalling aggression. Even if they are not read in this way, the inability to move the ears deprives

Above: The stumpy-legged 'Munchkin' has a restricted range of movements compared with normal cats

Far left: The flattened ears of the Scottish Fold

Left: The modern Persians generally have a lower activity profile than the heavy Maine Coon and Norwegian Forest cats or the light original Turkish and true Turkish Vans, or the newer Somalis and Balinese. Changes bred into the coat of modern Persians have also reduced self-grooming efficiency

Below: The controversial Sphynx seeks out warm surfaces and breeders say it normally prefers people to other cats

BREEDING FOR BEHAVIOUR

The Ragdoll is the first breed to have been specifically selected primarily on a behaviour characteristic. It was given its name and made into a breed purely due to one characteristic – that it would go limp on handling. It was termed 'the cushion cat' and was portrayed as 'the ideal kiddie cat' that could be wheeled about in a kiddie's cart without any risk to the children. When criticism was voiced that this extreme docility might not be in the best interests of the cat, breeders in Britain sought to achieve the attractive appearance without the limpness.

The American Ragdoll, as bred and handled by its originator Ann Baker, is an undeniably floppy cat. She believed this behaviour to be due to the involvement of the founding female, Josephine, in a road accident. Yet for an inheritable change it is more likely to have been a mutation, or as Desmond Morris has suggested that as both founding parents are from docile breeds – a Persian and a Birman – the docility may have been accentuated in this particular cross. Ann Baker believes that lines further from her original stock are more 'dilute'; certainly there does seem to be a range of docility, with some cats appearing little different from other Persian or Birman-type cats.

the cat of a major means of communication.

The American short-legged mutation called the 'Munchkin' has been selected purely as a novelty. It cannot jump and climb in the normal way, and lacks the lithe balanced movements of other cats. Its grooming behaviour is also restricted because of its short leg length. Normal behaviour is similarly affected in the Sphynx. Its lack of fur affects its ability to control its temperature; consequently, its free access to the outside world in cooler conditions should be curtailed.

4 Your Cat's Territory
Territory or Range?

Territory and range are words that are very often used interchangeably by cat owners, yet to a cat biologist they mean very different things. The territory is the area that a cat will defend against other cats, while the range is the area the cat normally inhabits. Territories are usually a bit smaller than the normal home range, and generally the larger the range the more differences are observable in its use. In general, the concepts of territory and home range vary according to the type of animal. The need to obtain food is common to all animals and the landscape dictates the availability of food, the best way for an animal to obtain it, and consequently its social structure. An open landscape encourages group grazing by herbivores, and hence a group predator (such as our dogs' ancestor, the wolf) is most efficient in hunting. In enclosed landscapes the reverse is true. The home range for lone hunting cats pursuing solitary prey is an individual affair, and in consequence so is its territory.

Home Range

I started to investigate feral cat behaviour and home ranges in the 1970s. Cats then were considered pets or pests: it was thought if they were not household animals then they must be starving. But when I weighed and measured cats from a number of colonies I found they were not that different healthwise to house cats.

I worked on numerous feral colony locations, but my main study, started in 1977, was sited in the heart of London in Fitzroy Square, at the foot of the Post Office Tower. The cats scavenged for food from waste sacks and bins around the square, but were fed a large proportion of their diet by kindly feeders. The male cats maintained home ranges of around 2ha (5 acres), while the females used ranges in and around the square of a little over 0.2ha (0.5 acre). The density of cats was around 12 per hectare (30 per acre).

I also carried out similar studies on the house cats of densely suburban areas of London, one of which – ironically for a cat-watching study – was called Barking! Even though house cats are domestic animals we have not, until the recent tendency to confine them in buildings, limited their ranges by barriers. Other household animals like dogs or gerbils are much more restricted, but cats pass through or over fences with ease. Yet they do not just wander randomly: they set their own territorial limits, keeping a watch over the cat next door. In one study area of nineteenth-century terraced housing and small gardens, with a density of 20 cats per hectare (50 per acre), I found the average neutered female to inhabit only 0.029ha (0.07 acre) – a much smaller range than the urban feral cats. The average neutered household tom had 0.11ha (0.27 acre), while intact toms averaged a bit more at 0.18ha (0.44 acre). In another site the cats were even more densely packed in. It seemed that our household moggies were prepared to accept smaller ranges than feral town cats.

How did country cats fare? A study of farm cats by David Macdonald and Peter Apps found queens to be using a range of about 6ha (14.75 acre), with the toms' ranges being much larger. The density was about 100 times lower than that of my city feral cats. This was remarkable, for although one group was feral and the other farm

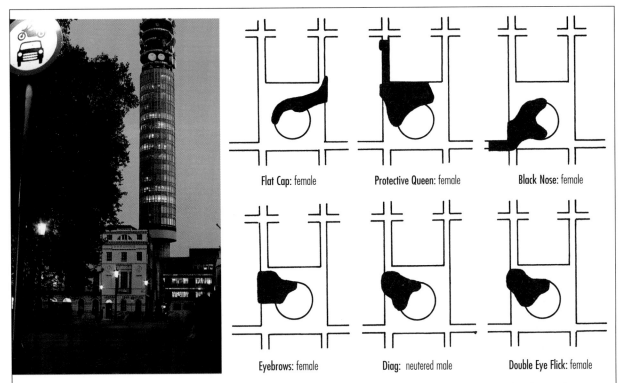

Flat Cap: female   **Protective Queen:** female   **Black Nose:** female

Eyebrows: female   **Diag:** neutered male   **Double Eye Flick:** female

The overlapping ranges of some of the cats from the world's first behaviourally-monitored urban centre feral cat colony at London's Fitzroy Square (pictured above left). The pattern found here was subsequently found to be typical in that females and males overlapped a common core area and that intact males had much larger ranges than the females. The ranges shown are after the colony had been neutered. This did not change the size of the female ranges, but the range of the male Diag shrank by nearly ten times to become the same as that of the females. Two other males following neutering moved away. Males commonly have a greater tendency to mobility

cats, in reality they both had additional feeding and both were free-ranging. Why should cats live at much higher densities in town?

Food at the Heart of Things

The main answer is simply the availability of food. The size of the home range of the cat is the size it needs to obtain its food. If food is plentiful it does not need a large range, but if food is scarce it does. Even for feral cats, the town provides richer pickings than the countryside. Despite its impressive armoury for catching and dispatching prey, the cat is a most competent scavenger and the rich pickings from tips and garbage sacks have allowed cities to house high numbers of feral cats. In addition, armies of dedicated feeders of feral cats have helped to support their numbers. It is fortunate for us that the cat is prepared to accept dead food, for otherwise the keeping of cats as pets would not be practical.

In various studies, toms were seen to have much larger ranges than queens. Toms are bigger but not

by much, yet male ranges have been found to be three to ten times larger than those of queens. This difference cannot therefore be due solely to food requirements. Radiotracking studies on wild cats, big and small, have demonstrated this same ratio.

The feral pattern provided me with the clue as to why this should be. In essence, the reason is social. The male ranges overlap those of the queens of what can be viewed as a group. The male buffers the queens of his group from other groups, and imparts social stability for a genetically related group of cats. The queen is the basic unit of the cat's pattern of land use. In any area she has the amount of land that is needed to support her; the tom defends a larger area within which the queens, which he is most likely to mate with, can rear their young more securely.

Where food is found by cats mainly in one place, such as around waste bins, where a feeder puts out food or at a rabbit warren, the cats have a common core area of overlap. Where food is sparse, the queens' ranges do not overlap as much.

HUMAN 'CATS'

In the lives of suburban cats the pattern of ranges becomes more cramped, because cats view us as both a food source and some sort of cat. Each house cat forms its own group with its owners, and has a common core area with them in the house. In identifying with its owners as its group, a queen behaves territorially towards a neighbouring queen as if it were a member of another feral group. Toms still overlap a number of queens' ranges, and will set the size of their range relative to these queens.

Females have only just a bit more territory than their own garden, while males have 3 to 10 times as much (indicated by the three different dotted and dashed lines).

Your Cat's Territory

Good Neighbours

When we have more than one cat, our homes become closer to a real cat group. When a new cat moves into an area there is a period of territorial adjustment, which can cause the stress problems seen on page 122; however, this usually settles after a while. As multi-cat households have become more common this has affected our cats' territories – for good as well as bad.

The more friendly you are with your neighbours, and the more you visit one another and go into each other's gardens, then the more tolerant the cats in each household will grow of each other. It is as if the perception of group size has grown, and this is not only true of multi-cat households.

In such situations, events can occur between neighbouring cats that would normally happen only within a group. For example, when prey is brought back into its garden by a hunting cat, then a neighbour's cat, whose presence has been

allowed in the hunter's garden, will take an interest, but unless it is particularly aggressive it is unlikely to intervene in the other cat's range. However, if the prey escapes, then not only will the cat who has lost the prey go back over sites in the garden where it had the prey and check them out, so too subsequently may the 'group adopted' neighbouring cat.

Your Cat's Range

Although your cat has a territory which it will defend, this is not an absolute size: it varies with the season, and even with which cat it is adjacent to. In the cold your cat may not wish to venture out much at all – sitting by the fire or a radiator

Above: The traditional centre of our own and our cats' ranges in winter – the hearth is home
Above right: The jungle retreat for one small cat – its garden shade spot

becomes its prime concern. It will have similar favoured sunning spots in which to snooze in your garden in the summer, as well as shady spots to avoid overheating. The cat will also have preferred latrine areas and patrolling and guarding points. Between all these it will have paths of greater use, and off these, paths of lesser use.

The cat's world inside the house, where we are in contact with it most often, is like the core area of overlapping ranges of a feral group. Depending on your use of the garden, the overlap with your cat will vary. Usually queens' ranges will approximate to your garden and a little more, while toms will generally wander further afield.

A transition occurs between the inside and the outside world when you open the door, or provide a cat flap. Because of this, flaps can become points of social tension. Generally however, their advantages easily outweigh their disadvantages. Allowing your cat its own access route in and out of your home is not just a modern idea. My own watermill home has cat holes in its ancient doors which allowed access for generations of working cats to carry out rodent control activities. The adjacent granary has the same. The mill stables also had cat holes to allow cats to wander around the working horses for the same duties.

Left and above: Cats patrolling their outside ranges at night, when they have the advantage over us, with their nocturnally adapted senses, such as being able to see and detect scents better (see pages 16–19)

Scent Messages

The cat is a fiercely territorial animal, but as most cats' territories are larger than they can see all at once, they are heavily dependent on leaving and interpreting scent messages. As a hunter, the cat has a finely attuned ability to discriminate between scents, and in the wooded landscape of its ancestors this ability was used for more than just catching prey. The world of the cat depends on the interpretation of various scents. Although some, like the spray of intact toms, are obvious to us, many are more subtle and can be overlooked by owners, yet they are most significant to cats.

As your cat goes through the cat flap it leaves grease from its coat on it, as if it has rubbed. Consequently, it will sniff the flap carefully from time to time to check if the scent on it is still its own, or if there has been an intruder. A strange scent can make the cat cautious and take longer to go out than at other times.

Rubbing

When you are out with your cat in the garden you will find it interesting to follow your cat quietly and watch what it is doing – but don't be too obvious! You may find its attention is drawn by scent to something on a plant tub or low wall at cat height. It will sniff carefully for perhaps five seconds, and may then rub directly onto where it has been sniffing. It may just rub with its cheeks along the side of the mouth, or it may continue to rub the side of its head and on until it is rubbing the back of its head against the scented object. It may then check out the smell of the object again, but normally more briefly. It may repeat the rubbing and checking again.

If the object is a leafless bush the cat will spend proportionally more time rubbing with the side of its mouth slightly open, but still carry out the other moves. If a stick is jutting out as a cat travels along a path, it may well sniff it and just rub it with its mouth slightly agape. It is also likely to brush against it with its body, a move which is not normally noticed by us.

'Allorubbing' creates group scent between cats in multi-cat households and between individuals in feral groups, and like much in cat behaviour it has more than one function: cats transfer scent and gain a mutual group identity at the same time as encouraging group bonding. At the Fitzroy Square colony, I found that cats would keep a normal minimal inter-cat distance of 30–60cm (1–2ft) most of the time (aside from feeding). However, when anticipating feeding, cats would rub after tail raising.

GROUP SCENT

Although cats actually have very few sweat glands in their coat, they do have sebaceous glands to protect the hair and these are also a source of scent. The glands can be found on the lips and chin, the top of the head and also along the top of the tail. When we stroke a cat or it rubs itself against our legs, we pick up these scents ourselves. Consequently, we then have a group scent identification. Cats can react by hissing at someone if that person has previously stroked another cat that the first cat regards as territorially antagonistic.

Far left: American homestead cat head-rubbing a logpile

Left: Chinning and rubbing the gatepost on your arrival

Below: All may not be what it seems! Anxious neutered toms and queens, emboldened by their owner's presence, may well 'air spray' when leaving the house. This means that they go through all the movements of spraying without spraying

Chinning

Cats rub with the side of the mouth for they have large sebaceous scent glands along their lips and chin. This sort of move is called 'chinning', and is usually a specific reaction to scent. You may find your cat crouched low to the ground, extending its chin and rubbing it on the ground. It may be attracted to residual scent from another cat's anal scent glands, or to scent left where you have been standing, or to some other significant scent.

Usually the cat will carry out the move in a straightforward way. At other times it acts more obsessively, determinedly sniffing and rubbing. Just as with its response to catnip (see page 43), the cat may make the ground quite wet with its saliva. Also, again in the same way as with catnip, once its chinning episode is finished the cat will show little further interest in the spot.

Spraying

When patrolling his territory an intact tom will reverse up to an elevated object, lift his rump high by raising one of his legs to stand tall, and, with tail erect and quivering, spray a highly pungent stream of liquid. He will do this regularly. Among farm cats, it has been found that toms will spray more often when there is a queen in oestrus nearby. Although queens can also spray, they do so far less often.

The most frequent rate of sprays found by Peter Apps in a tom patrolling farmland was 63 per hour. At this rate it is clearly not to empty his bladder, and patrolling spray volumes are usually quite small. Corbett studied cats on the island of North Uist, and observed that while patrolling the rabbit burrows they normally sprayed every 5½ minutes.

Spraying, and then reading its own scent and not another, seems to give a cat confidence of territorial ownership. The frequency of spraying can increase in an area of territorial dispute. When a neutered tom in a contested area gains confidence he may well spray vertically in the same way as an intact tom, but it will lack the all-invasive pungent odour. As key crossing points, it may be that doors with cat flaps are specifically sprayed in territorial conflict. (See page 124 for problems with spraying.)

Scent Messages

The role of spraying in increasing territorial confidence is further seen in the location of patrol spraying. This is carried out in the areas of main use, such as hunting areas, while there are fewer sprayed points at the believed edge of the home range: this highlights the difference between territory and home range. Cats do not behave as if they are doing guard duty circuits; rather, it is a matter of confidence and usage. That spraying is in large part for territory declaration is evidenced by the fact that toms spend more time investigating spray from other intact toms of unknown origin than from those of their own or an adjacent group.

Territorial declaration – and exercise!

Clawing

Cats claw into trees and other wooden objects, dragging their claws vertically on outstretched paws through the wood to sharpen them by breaking off shards of keratin. The wood has to be of the right texture to pull through for the cat to use it. As well as this, cats will also claw horizontally on branches, fence rails, tree roots, matting and furniture. Cats sweat from their paws, not their coats, and consequently leave faint scent messages when they claw mark, which would be overlooked were it not for the visible scratch marks. Dennis Turner has suggested that as it happens in front of other cats more than when cats are alone, the visual act may have a dominance significance. But it may be mutual territorial assurance as cats do it similarly in front of owners. (See page 125 for problems with claw damage.)

BURYING DROPPINGS

Cats are famous for burying their droppings, yet when there is a territorial dispute between male house cats, both intact and neutered toms will leave droppings prominently sited and not buried. Studies have also found that farm cats that bury faeces around the farm may leave them exposed elsewhere. House cats prefer to bury their faeces on the edges of their range, which often places them in the neighbour's garden – not always appreciated by the neighbour!

Cats are quite sensitive souls, and need to feel confident even to go to the loo. They become concerned if other cats are about, for while using a dug latrine they are vulnerable to intimidation. Consequently you may find that, particularly in the winter, as you come home or go out into the garden you are accompanied, as your presence gives protection. (See page 122 for problems with fouling.)

Cats and Catnip

If you are trying to grow catnip (*Nepeta cataria*) in your herb garden, you will not need telling of its powerful attraction for some cats. The local cats will have been drawn to it by its scent, and will have repeatedly rolled on, sniffed and bitten the plant. The volatile oil in the leaves and stems stimulates the olfactory bulbs at the front of the cerebral cortex and then produces altered states in the cat.

Taking advantage of the herb's powerful attraction for some cats, cat toy manufacturers often put some of the dried herb or extract in their products. In fact, cats are often affected more strongly by the dried herb than by the green plant.

'If you set it, the cats will eat it; if you sow it, the cats don't know it.' This old gardening saying is revealing of the cat's responsiveness to the plant: transplanted plants are quickly sought out by cats, while those grown from seed are often untroubled unless they are handled. The logic of the rhyme is based upon the fact that bruising the leaves and stems releases more of the scent which attracts the cats.

Catnip has square stems, which shows that it is a member of the mint family, and attractive white flowers, and grows wild in America and Europe. In Britain, it is a native plant and is found growing mainly in the east of England on dry chalky soils.

The catmint that is more commonly grown in gardens as an attractive lavender-flowered edging plant is *Nepeta mussinii*. Cats are not drawn as strongly to this form, and larger *N. m.* 'Six Hills Giant', as to the native one, as the scent is far weaker. If you wish to grow the stronger-scented native form for your cats without having it destroyed, the trick is to grow it through other plants which will support it.

The domestic cat is not alone in its response to catnip, for even lions will respond in a similar way. The behaviour in both sexes resembles that of the female rolling oestrus and post-copulation behaviour. It is also displayed by both neutered and intact animals. It seems probable that catnip induces this behaviour because its smell resembles the female cat's sexual odour.

Not all cats are affected by catnip. The ability to react seems to be inherited as a dominant gene and does not show until the cats reach sexual maturity. There may be a breed linkage, as Siamese seem less affected. The desire to roll in catnip may arise from changes in skin sensitivity or glandular secretions that may be caused by oestrogen.

Cat owners are sometimes concerned that if the mere smell of catnip induces a strong response, eating it will make the cat worse, and with access to the plant or dried material cats do readily consume the herb. However, it has recently been reported that when taken internally catnip actually has a calming effect on cats, in contrast to the states of arousal produced by its scent.

Whether or not a cat responds to catnip depends on whether it has inherited the necessary receptiveness, and surveys have put the incidence of this at around 50 per cent of cats. The most common response is to show interest by sniffing, then often to lick it or eat it. Some cats, particularly older ones, will drool noticeably. Quite often, these reactions may be followed by a small amount of chinning (see page 41). These animals often lose interest soon afterwards. However, those showing a keen focused interest will endeavour to pull the herb to their head with their paws while rubbing against it, even if not they do not exhibit head rolling. With this group, if an attempt is made to pull the catnip away the cat may well strike out at the handler with claws outstretched. This behaviour will occur even with cats that have no history of striking out at people. This type of response can also occur with cats that enter a heightened state from inhaling powdered catnip, and owners need to be aware of this risk when giving their cats catnip materials. However, this state will pass off within seconds of the catnip being removed. Shy cats often exhibit an inhibited response, whereby all they do is sniff the catnip with some interest, and even keen cats will eventually reach a satiation point.

Due to the range of possible responses, it is not easy to draw an absolute line stating when kittens will begin to show an interest in catnip. Young kittens do not respond, but post-weaned kittens around 9 to 12 weeks old will begin to chew on the stems of the dry plant and sniff at it with interest.

5 Hunting Behaviour
Hunting Tactics

House cats have been recorded as hunting for an average of 15–25 per cent of the 24-hour day (with some hardly at all). Rural feral cats may spend up to 45 per cent of their time on extended hunting forays and are more active in the summer than the winter. So well known is the cat's love of comfort, and its normal dislike of becoming wet, that most people assume that cats will not want to be out on a wet night. This may be true of the pampered lap cat, but the skilled hunting cat knows the value of a really wet night, and will venture out after only a slight hesitation. I have noted that on such nights prey size may be larger, and the success rate better as the noise of the heavy rain falling on the ground and leaves masks the hunter's stealthy approach.

The Stalk

We usually picture the cat's hunting behaviour as the classic sequence of a stalk, followed by capture leading to the kill. But although frequently used to catch birds on the ground, this sequence is not appropriate in every situation.

Most successful hunts involve small mammals as prey, and while a stalking run is often used, skilled hunting cats are always alert to opportunities even when just strolling around or apparently snoozing. Carnivorous shrews must eat every two hours, day and night, winter and summer, and become so focused on their foraging and hunting that they are oblivious to their rustling among fallen leaves. In contrast, herbivorous voles make trails above ground but underneath the mat of grass, and so are out of sight. In both cases, when the cat hears a movement the potential prey may be less than a metre away. As sound sources are harder to detect when moving, it will pinpoint the source of the sound by sitting still and tilting its pinnae towards the noise.

Unaware of the danger, the prey may wander out in front of the cat into the open, but this may inhibit the cat, who will often not dash and grab the unwary prey but just paw at it tentatively.

House cats are often unperturbed if their prey dashes behind vegetation or some object, and will spend some time trying to dislodge it from its position. If the animal has gone into a burrow, the cat will try to extract the prey with its paws extended. If successful, it will again resort to tentative patting. The cat has every justification for being tentative, as a vole or shrew will certainly threaten the cat with its teeth.

The Capture – A Dazing Pounce

If a vole manages to evade capture or skips back into the grass, the cat will relocate its position with its ears and then use its vole-dazing pounce. To do this, it rises up on its hind legs and brings its weight down through stiffly held front legs onto

Left: Poised stationary, the ears have pinpointed the prey, and the eyes have focused on where to pounce

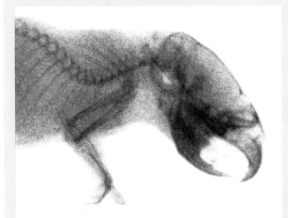

THE KILL: THE NAPE BITE

Cats deliver the nape bite in its most precise form. While some species of small wild cat deliver a nape bite at once to their prey, the cat may go in with paws first if safety reasons require the prey to be dazed. With domestic cats, the prey is generally dispatched with a nape bite after it has been dazed, although on some occasions the capture holding bite is delivered in such a direct way as to kill the prey.

From extensive observations Paul Leyhausen believed that cats hit the cervical spinal cord in a very high percentage of cases. He took as signs of fatal injury that the prey animal's eyes bulged, the rump and limbs convulsed and the tail stretched stiffly. The cat's canine teeth are particularly well served with mechano-receptors, so when the teeth make contact with anything such as bone when moving across muscle they have the potential to adjust through nerve feedback, inserting a tooth between the vertebrae like a wedge and severing the spinal cord. With Alan Hatch I examined by X-ray numerous small mammals killed by cats, and found no vertebrae damage.

the prey. (Foxes make a near-identical pounce.) This stunning blow can knock the air out of the vole and cause it to give an involuntary squeak. When the grass is tall and tangled, preventing a steady approach and possibly allowing the prey to disappear, the experienced cat may make a distant leap. It will ease its body back without moving its feet and then let fly in a high, curving arc to land on the prey.

In hunting a bird less chances are taken, but a cat still does not always employ the classic stalk method. If a bird is on a low branch, an experienced cat can sprint, leap and grab the bird in one smooth action.

Predator and Prey

Prey Defence

Prey animals are fully capable of self defence and cat behaviour takes account of this. Between the initial catch and the death bite, birds can escape and fly away. Moles will turn defensively onto their backs ready to give a formidable bite. Rats, mice, voles and shrews too will bite readily, so the cat is cautious in its approach. A study by Leyhausen found that if a rodent squeaked loudly even when it was touched only lightly, the cat was highly likely to leave it alone. However, if the cat was hungry – that was another matter!

Playing with Prey

When engaged with prey, cats may be seen to 'play' with it. Even in large cats, this 'inhibited play' seems to stem from a fear of injury by the prey. One of the functions of so-called 'playing with prey' is to tire the prey to make it more vulnerable to a neck bite. This is particularly effective with shrews. As carnivores themselves they are ready to bite the cat's muzzle by turning over, but when first captured and released they run fast rather than remaining still. On capturing and recapturing them the cat has to act quickly, but is wary of the bite. Yet as the shrew wearies

PREY CATCHING – NATURAL OR LEARNED?

Electrostimulation studies of cats in the 1940s, 1950s and 1960s showed that even cats that do not normally catch prey and had not learned to do so could be triggered into killing prey, although the effect was temporary. This suggests that seizing and killing behaviour is a natural instinct within all cats. However, these stimulated kills lacked sophistication, unlike those of cats that had learned through kittenhood about catching and killing prey. These cats learned to dispatch prey with a correct neck bite after ensuring that it was sufficiently subdued and dazed by paw battering. This expertise and fineness of judgement seems to be a behaviour that is learned rather than innate and as such ensures survival.

its running is reduced, although it is ready to defend itself to the last.(See pages 48–53 for a detailed account of a cat 'playing' with two different types of prey.)

Bringing Home the Catch

Many people are perturbed that their cat brings prey back home: some go further and suggest that such behaviour is devastating wildlife populations. In reality that is far from the case, but why is it that cats bring back prey either into the home or just outside?

In part, the cat is bringing food back to its hopeless kittens (us) who have not yet learned how to hunt. It also has the positive effect of demonstrating to other cats of its group that this is a good area in which to hunt. Primarily, however, this behaviour has a very practical role in achieving hunting success.

When a cat initially grabs the prey with its mouth, it catches any part of the prey it can. To kill the prey it needs to use a neck bite but, unlike a pack of dogs, it cannot release its grip while another cat holds the victim. The cat therefore needs to release its prey and then bite it at the neck, but during that critical move the prey may escape. Consequently, two pieces of behaviour have developed to assist in the repositioning.

The cat is reluctant to release the prey where it was captured, as this might allow it to escape back into terrain the prey knows intimately, giving it the advantage. In contrast, if the cat takes the prey back into the core area of its own range, which is the area it knows like the back of its paw, it is likely to be more successful in recapture when trying to deliver the death bite. If the prey does escape, it is still within the cat's area and liable to recapture. Unfortunately for squeamish owners, the core area of a house cat's range equates to our common area with the cat!

Dazing the Prey

The initial catching of the prey and bringing it home is only part of the story – the cat still has to dispatch it. Most critics of cats, and indeed many

THE EFFECTS
OF SEX HORMONES

There is some evidence that sex hormone levels affect the readiness of a cat to catch prey. In the female cat, high levels of circulating folliculin increase this readiness, while oestradiol produced by the corpus luteum in the ovaries decreases it. In the male cat, higher levels of testosterone increase its readiness to hunt. It can thus increase its range size and find more prey. In addition, the larger the size of the home range, the more likely it is to overlap more female ranges, and so the tom can sire more offspring (see Chapter 4).

the cat has excellent vision, due to the flatness of its face, as it approaches within the last few centimetres the area around its mouth becomes virtually invisible. The whiskers may come towards the prey when required, but this does not overcome the problem.

In consequence, the dazing of prey, which is commonly viewed as 'playing', is essential for the cat. Infections arising from bites from prey could be fatal and are best avoided. Once the prey is apparently dazed, the cat needs to check that it is actually dazed before risking contact. So the cat will sit and look around and away from the prey with apparent disinterest. If the prey is not dazed but watching for its moment, it will use this time and flee. The cat will then whirl around upon the prey and the chase will resume, until the prey is sufficiently dazed or has escaped.

owners, assume that 'play' with prey is just an unnecessary exercise in prolonged cruelty. However, the cat actually has a problem in that even though the prey is now on the cat's own patch and restrained, the cat still has to release it and apply a neck bite. As lone hunters they have short muzzles to enable stronger bites, so although

Even after prey has been killed, cats will usually bat it a few times to make sure. In a domestic cat that rarely kills this may become prolonged, then becoming recognisable play

During release and recapture ('playing with prey'), this picture illustrates what happens for most of the time. The bird remains motionless while the cat appears disinterested, looking away, in an attempt to convince the bird that it is not watching it

The young robin is demonstrating the typical defensive ploy of remaining stock still, which allows the cat to check it out at very close range by sniffing. The cat's behaviour would be different if this were a small mammal: when the bird moves off, the cat must pursue it instantly or it will fly away; when a mammal runs around, the cat can allow it more freedom of movement as it is easier to recapture

Playing with Prey

As there are both distinctions and similarities between a cat 'playing' with a song bird and 'playing' with a small mammal, on these pages I have set out a detailed comparison of two such typical play and prey sessions. These were filmed as they happened, from captures of wild prey made by the same house cat without human intervention, and the film was then analysed to determine the exact sequence and timing of events.

Cat with Bird

The bird captured by this cat was a young robin and, as happens quite commonly, it made good its escape during the release and recapture events of 'play'.

Initial Play

With the bird held just as it has caught it, the cat takes its prey to its own garden, lies down sphinx-like and releases the bird. The cat then taps the bird with its right forepaw, and the bird takes off. Within 1 second the cat brings down the bird with its forepaws and holds it under its forelegs and head. It then eases back, keeping a paw on the bird. After sniffing the air, the cat walks away for a metre or so, where it lies down again. The bird remains motionless, so the cat circles back and once again lies down sphinx-like. Then, again using its right forepaw, it taps the bird.

The bird flies away for a metre or so, but the cat again brings it down, this time within 3 seconds, once more holding it down under its front paws and head. The cat then carries the bird back to exactly the same spot again. What is particularly remarkable is that the cat has not just brought the prey back to its 'own patch', but chooses one particular spot at which to release its prey each time in the sequence.

Once again, the bird takes off and this time evades capture for a distance of about 5m (15ft).

However, slow-motion analysis of what takes only 8 seconds shows that the cat is more in control than appears to be the case at normal speed.

Release and Recapture: 1

From the point of release of the bird, the cat has both front paws arched above its head in just ¼ second. Within a further ¹⁄₂₅ second it has both launched into a spring from its back paws, and brought its front paws together on the bird. Using its paws and head, it then traps and pulls down the bird. The cat's leap has taken it the same distance as its outstretched length, and it lands first on its back paws. To land next on its front paws it has to let go of the bird, but its swinging momentum has pulled down the bird, which hits the ground.

The bird now flies off at an angle of 90°, but the cat follows, keeping its eyes fixed firmly on its prey. The cat again springs after the bird and swings its paws to grab it, once again pulling it down to the ground. This is the adult use of one of the four key types of play move in kittens, which seems so exaggerated when used in early play patterns. However, in the recapture hunt of 'playing with prey' so much happens so fast, and is co-ordinated so precisely, that the owner will usually be unaware that the adult cat is making this particular move. The sequence of moves from the point at which the cat re-released the bird at its favoured spot, brought it down to the ground twice and covered a distance of 5m (15ft) has taken less than 2 seconds! Film analysis is essential to follow such speedy pursuit.

Escape Opportunities

The bird now slips out and forces another change of direction, but is unable to gain much height before the cat runs it down within a few paces, trapping it below its chin and front paws. The cat repeats its movement, but so does the bird! The cat runs forward, trapping the bird under its chest and right foreleg, but despite a containment move the bird slips out.

The cat follows, once again springing into the air with its front paws up and bringing down the bird. The cat then runs onto it and holds it to the ground with its mouth, but loosely enough for the bird to be flapping a wing. During this sequence of the repeated bringing down of the bird, there have been a number of occasions on which it nearly escaped, with only the skill of the hunter keeping it contained.

When recapturing the bird during 'play', the cat restrains it until it is sure that the bird will remain still, but it does so with an extended paw and keeps its head well back. (When the cat originally caught the bird it was feeding on a worm, which can be seen in this picture on the bird's back.)

The Second Hunter

The event that follows is not unusual, for individual cats do not operate in a vacuum, and one of the risks of bringing prey home is that it may be taken from you by either your owner or another cat. In this instance, the other cat from the hunter's home is attracted by the flapping of the bird and approaches to within 60cm (2ft). This distraction allows the bird to flutter away for a metre or so, and the second cat follows it. This inhibits the first cat, who steps away. The new cat walks up to the bird and sniffs its back. The bird remains stock still during this contact-sniffing, which takes a total of 7 seconds. This 'nerves of steel' behaviour by the prey is normal during prolonged periods of the recapture hunt, and has the inhibitory effect of imposing caution upon the cat. In this instance, the second cat is not a skilled hunter, unlike the first. While it could easily give a nape bite while sniffing the back of the bird, it does not do so. Instead, it walks a little way away and then returns to sniff cautiously, some 7.5cm (3in) from the bird. The second cat then sits back down, sniffs the air and the ground, moves right around the bird and sits down again.

After 3 minutes, the first cat is back in position. It then contact-sniffs the bird, looks around sniffing, re-contact-sniffs the bird, and then looks around sniffing once more. The bird has not moved at all in 3½ minutes, so the experienced first cat makes what might seem a strange move. It walks 3m (10ft) away from the bird, back to the exact spot where it first positioned it, leaving the bird standing rigid behind it. Such apparent indifference on the part of a practised hunting house cat is not at all uncommon. With a static prey like this the cat seems to follow one of two options: it either re-taps to test the prey's degree of dazedness, or it waits with apparent indifference to allow the prey time to 'unfreeze' and attempt to flee, which then triggers the cat to recapture it.

However, such actions leave the prey vulnerable to piracy by another cat, and the second cat returns to investigate. Again, it contact-sniffs the bird. Normally, cats do not go to face their prey by choice, and this cat shows its inexperience by moving in front of the bird. It then sits down so close beside the bird that it is almost on top of it. However, this cat too shows apparent indifference by looking around and away from the bird; then, after a few minutes it moves more correctly behind it. After this the cat moves back, while the original hunter is still lying down 3m (10ft) away.

The bird is still motionless but has its eyes open, and now, after 4 minutes 43 seconds, for the first time it moves its head to look around. When it has clocked up 5 minutes, the young robin looks around a little more, but cautiously. This seems to be what the hunting cat is waiting for, and it goes to the bird and sniffs it, then sits down. Like the other cat, it sniffs the air and looks around. However, although it seems to be merely looking around, in slow motion it is apparent that the cat looks directly at the bird in passing. The bird

THE KILL: LARGE AND SMALL

In general, big cats kill smaller prey by a nape bite in the same way as small cats, but for large prey they use an asphyxiation clamp bite at the throat of the animal. This is due partly to the size and thickness of the neck of some large prey, but also because antlers or horns would make a neck bite dangerous to the cat.

The nape of the neck of the prey is a key stimulus to the cat to orientate its killing bite. This is so important that in man-eating areas of tiger country, such as the Sundarbans or parts of the western Terai in India, those at risk such as fishermen and wood gatherers who expose a low neck profile may wear a shield to disguise their nape.

By taking large prey big cats do not have to kill so frequently; small cats catching small prey may have to kill a number each night. The prey of small cats is often nocturnal, so the cats' hunting timetable corresponds. The prey of the big cats is much more mixed, and so is their timetable. One huge advantage for big cats in taking large prey is that they have far fewer competing predators than do small cats. However, by killing prey larger than they can consume at one time, they are presented with many competitive scavengers. As this distinction is so significant in relation to hunting success, minimising the loss is critical for the survival of big cats. The tiger relies on cover, and the leopard takes to the trees. The group feeding of lions, which are plains animals, is one of the single greatest advantages of living in a pride, for by all eating at one time the possibility of food loss is greatly reduced.

looks at the cat. The cat continues to look around, even looking right away, yet about ten times its eyes check the bird since the cat repositioned, and its ears turn towards the bird. The bird has not moved at all for 7 minutes, so as the cats' 'disinterested' glances pass back and forth they are directed less and less towards the bird.

Escape and Recapture: 2

This strategy works for the cat, for after a full 7 minutes the bird senses it can escape, suddenly looks directly at the cat, turns its body and takes off, with the cat in hot pursuit. The bird lands 2.5m (8ft) up on an ivy-clad wall, 6m (20ft) away. The cat runs up the wall, and comes back down with the bird in its mouth. Escape to recapture has taken only 10 seconds. Remarkably, the cat once again positions the bird back in exactly the same place that it first put it. No sooner has the cat placed the bird down than it looks away! However, it looks back down at the bird just as quickly.

The initial glance away by the cat is to check its own security, and having looked at the bird it sniffs the air and gives a quick lick in a move that is characteristic at such times and seems to be a 'fast Flehmen' without gaping (see Chapter 1). The cat looks around again, and then licks in the same way once more. At this time the cat's pupils are noticeably narrowed while it continues to look around and sniff. After 30 seconds of this, the cat suddenly gets up and moves back 1m (3ft) behind

The onlooker watches the cat in the arena of action. There is always a risk in bringing prey 'back home' in that the hunter could lose it to another cat

the bird. It settles down, and continues to look around and sniff the air.

Escape!

After a few minutes of the cat's apparent absence the bird has still not responded by attempting another escape, so the cat moves back to re-investigate. It contact-sniffs the bird briefly, then again sits and looks around as another quick security check. The cat then looks directly at the bird and puts its left forepaw on the bird's tail. It goes down sphinx-like, but keeps its head well back and its forepaw out in front on the bird. The cat taps the bird's back, and it responds by flying off.

Within less than ¼ second the cat has one of its paws holding the bird into its open mouth as the bird flies. Springing after the bird, the cat pulls it down onto the ground, less than ½ second after the bird took off. But as the bird bounces on the ground – just as it did before it changed direction by 90° earlier on – this time the strategy works, for despite the cat's skill and speed the bird flies to freedom.

Studying this whole 14 minute sequence in detail shows us that the bird made at least eight escape moves before it finally flew off. It also demonstrates that the cat technique for capturing birds is a definite spring up and then a pulling down movement with the front paws.

The cat looks down with its ear pinnae focused intently on the vole, to fix its position before pouncing on it. The cat's technique is to tap the vole when in the open with a paw, to probe after it in retreat crevices and as here to 'vole pounce' when the vole is under grass

Cat and Small Mammal

The same cat brought a field vole back to its garden and released it in a corner made by a brick and stone step. As small mammals invariably like the security of edges, this was a good strategic place to release the vole so that it could be recaptured and confined.

Initial Play

The cat sits, boxing in the vole by its position, but instead of studying the prey continually it looks around, with occasional sniffs of the air. After 1 minute the vole has not moved, so the cat rolls over on its back and taps the vole. Having lowered its height, the cat is rewarded by the vole moving, strangely towards the cat. The cat rolls onto its side for more control, with its face just a few centimetres from the vole. The vole remains motionless, so the cat turns its back on the animal. This has no effect, so the cat sits up and taps the vole with its right forepaw. The vole instantly moves, but within 1 second the cat puts its paw in front of it to block its escape, but moves the paw back from a threatened bite. The vole almost makes it to a gap in the bricks, but the cat uses its other paw and the vole turns back. As it runs, the cat focuses its ears on the prey and leans forward. The vole's move out and return to its corner has taken only 30 seconds, and with it safely cornered the cat checks out the vole's route by sniffing.

Advance and Retreat

The cat attempts to pat the vole twice with its left forepaw, but gives a flinch reaction as if bitten. After 30 seconds of inaction, the cat moves up onto the step above the vole, from which it cannot be seen so easily, and waits. The cat swings a paw towards the vole, but this is a inhibited move, and the cat ends up by giving the leg a displacement-activity lick.

The cat then brings a paw down heavily on the vole, causing it to squeak in response and move slightly. The cat puts out a tentative paw towards the vole but withdraws it; it then lies down on the step above the vole, turning its head upside down nearer to the vole. The vole stands on its hind legs and tries to bite the cat's ear as it passes. The cat then taps the vole and the vole returns to the corner.

The cat steps back down to its original position, sniffs close to the vole, then taps the vole with its right forepaw. This time the vole runs forward and into the gap in the bricks that the cat had previously blocked. The cat pokes its paw after it, and with a squeak the vole turns around. The cat puts its paw forward tentatively, withdraws it and then brings its nose down towards the vole, which again threatens the cat with an open mouth. The cat pulls its head back smartly. The cat then taps the vole to pull it out of the hole, knocking the vole a little off balance, and it squeaks again. The cat then taps the rear end of the vole gently, sending it running back to its corner again. The vole overshoots, so the cat directs it back with a gentle tap of its right forepaw. The vole squeaks and the cat gets up onto the step once again. It is now about 4½ minutes since the prey was brought back to the garden. Looking away, the cat yawns.

In and Out of Hiding

After a few seconds, the cat returns to find the vole is about 30cm (1ft) out of the corner and taps it, which sends it back to the corner. After a further minute of similar movements, the vole reaches another crevice, leading to a hollow behind two bricks. Lying down, the cat pushes its paw in, but then withdraws it. The cat then pushes its head into the hole, the vole squeaks and the cat rapidly withdraws. The cat circles around and puts its paw into the hollow but pulls its paw out rapidly.

The cat sniffs the hole and then circles back to the other entrance, and then puts its paw in, tapping vigorously a few times. It then runs back round to the other exit, where it looks in. The cat goes back once again to the other end, then moves back and forth repeatedly from one end to the other. Using a hooking action, the cat eventually gets the vole out, positions itself between the hole and the vole, and taps the vole again. The vole sets off across the garden earth, leaving cover behind. After a metre or so it stops, but so does the cat.

Self Defence

After 20 seconds without movement the cat gets up and taps the vole with its paw. The vole spins around with open mouth and tries to bite the paw. The cat sniffs the vole, withdraws its head and taps the animal again with a paw. In response, the vole stands up on its hind legs and tries to bite the paw again. Remarkably, it runs

directly under the cat, who steps back.

The vole goes into the edge of the long grass of the lawn. The cat tentatively taps at the vole. It then uses vigorous paw movements, concerned that the vole is escaping. The vole squeaks four times and the cat becomes cautious. After a series of tentative prods into the grass, resulting in corresponding squeaks and the cat reacting from

This remarkable picture shows the vole spinning around and threatening to bite the cat's paw, and is the key to understanding why cats 'play' with their prey. The cat's approach is inhibited by fear of the prey's defence, making it truly a 'scaredy cat'! (The vole threatened the cat at least 6 times in 11 minutes.)

one as if it has been bitten, the cat stops, sits up and makes a vole pounce. This elicits a squeak, so the cat reverts to the prodding and then stops again.

Escape!

The cat sits looking around for 30 seconds, sniffs the air keenly a number of times and then suddenly, fixing its pinnae on the spot, leaps up and makes a more vigorous vole pounce. The vole squeaks again, so the cat taps it tentatively once more. The cat looks around and sniffs the air, but the vole makes good its escape into the grass.

During the 11 minute sequence, in order to escape the vole's aim was to find cover, which it achieved initially behind the bricks, and then successfully back in the long grass.

6 Maternal Behaviour

The Mating Cycle

Today the great majority of owners have their cats neutered, so the mysteries of cats' mating behaviour are encountered less than in the past, although you may still hear the caterwauling of embattled males when others cross their territories to find a queen on heat. Remarkably, despite domestication and increased form diversity through breeding, cat mating behaviour has been modified little over thousands of years. One major recent change is that the warm, light conditions of a breeder's cattery or house enhance the number of cycles of sexual responsiveness in the queens each year.

The Fertility Queen

The queen goes through a very distinct cycle of behaviour which is linked to her hormone cycle. The shortening day lengths cause most cats to be anoestrus (sexually inactive) in October, November and December in the temperate northern hemisphere (and counterbalancing months in the southern). During this period there is minimal mating. The queen also has sexually quiescent periods between oestrus cycles of activity within the year. These shorter anoestrus periods are termed dioestrus.

Any tom attempting to mount the queen when she is not in oestrus will be dismissed by her clamping down her tail, and possibly striking at him. As she goes into the period of pro-oestrus, she starts to become demonstrative and will rub more around objects with her head. This is not a long period, lasting only one to three days, and it is typically noted by owners as an increase in their cat's friendliness. During this period a queen can develop a preference for a particular male that courts her by spending time sitting near her. At the same time, some males may try to mate with her but will be rejected. As the queen moves towards full oestrus her movements become progressively more demonstrative, she will open and close her paws, and she will begin rolling on her back and from side to side. She will also start calling – a sound that breeders of Siamese will know better than anyone!

As the queen's ovaries become ready to ovulate she enters oestrus. She continues to call and roll, and will also push the back of her head firmly on the ground, thereby leaving sebaceous scents around. The queen is most receptive to male advances on the third to fourth days of oestrus. If she is mated, her oestrus ceases within 24 hours under the influence of the progesterone produced from the corpus luteum in the ovaries. However, if she is not mated, she can remain in oestrus for a fortnight, although most noticeably only for around a week.

The overall average cycle without mating is about three weeks, but it is very variable, particularly across the pedigree breeds. In Siamese the dioestrus is very brief, while some Persians have less frequent oestrus. As Siamese queens commonly have long periods of oestrus calling behaviour, and calling itself can be more frequent, it can reach the point where it seems almost continuous! You have to be particularly dedicated to be a breeder of Siamese.

Why Cats Are Promiscuous

Observations of mating behaviour led our medieval ancestors to castigate the cat as promiscuous, wanton and therefore wicked. These judgements were based on the fact that cats mate repeatedly over a long period; the queen mates with a number of toms; and queens cry out not just with copulation but particularly on withdrawal.

However, these behaviours are due to territorial requirements and anatomical and physiological adaptations.

Cats are intensely territorial. This favours resident group males in mating, giving them preferential selection by the queen. It grants social stability, but carries the dangerous risk of inbreeding. Cats therefore require, and have, a mechanism that works to prevent this.

In most mammals, when an ovum is ripe in an ovary, it is shed spontaneously. In cats and a few other mammals that are induced ovulators, this does not occur. They need a triggering device to cause the egg to be released, and in the case of the cat this is provided by the anatomy of the penis, which is covered in backward-pointing spines that are longer on more sexually mature toms.

In most mammals, ovulation is independent of sexual activity, but in cats copulation is needed to stimulate ovulation. The raking of the spines of the penis inside the queen stimulates the release of the egg, and understandably causes her to scream. At the first matings, the spines initiate the release of the egg, which occurs 24 hours later. The egg then travels down the fallopian tubes until it reaches the point at which it can be fertilised.

Buying Time

This system increases the chances of fertilisation and also prevents problems with inbreeding. Generally, the cat's wild ancestors lived at much lower densities, separated by their territorial patterns. The initial matings stimulated the release of an egg and bought time for other toms to reach the mating area. Today, with urban feral cats and house cats living at higher densities, even more toms will gather and this has increased genetic diversity still further.

However, for this system to work the queen must be able to mate repeatedly over a long time. Seen in a feral group the system makes sense, but at a breeders the one tom is initially very keen, but as mating follows mating he tires and the gaps between matings widen, despite the queen becoming increasingly enamoured.

There is a possible flaw in this system: if the resident males could be displaced completely by interlopers, why should they continue to defend their larger territory, thus supporting another male's offspring? Without genetic flux the group's survival would be at risk, and consequently the male's genetic lineage. This problem is solved by multiple mating, which allows for multiple fathers although most of the offspring are likely to be those sired by the group males.

Most cats, like this Abyssinian, make excellent mothers

Mating

Although the queen has an oestrus cycle, and there is also a seasonal cycle, the event of mating (which can average around 40 sexual acts in 24 hours), also has a repeating cycle in itself. Between matings at a breeding establishment where only one stud male has access to the queen, there is a relatively prolonged inactive period of 5 to 15 minutes while both rest. The male, however, is not indifferent and sits close by the queen, ready for re-mating.

Courtship

As the tom judges it is time to re-mate he sits up, leaning forward encouragingly, and may make a quiet 'chirrup' call. When the queen is ready she will move forward from her resting position, crawling forward on lowered front legs and so walking her rump into the air to adopt a full lordosis position, at the same time looking at the tom to convey that her movement is for him. The raised and held-over tail of the female releases a waft of sexual scents to the male, and the readiness of her posture gains his interest.

During much of their day-to-day interaction, cats studiously avoid staring directly at each other, and will observe tangentially while apparently looking elsewhere. Yet during the repeating cycle of mating the queen and tom will spend a significant amount of time looking directly at each other. This occurs during the rest period and particularly during the mutual courtship.

However, the tom does not just leap upon the queen even now. While looking at him, she may blink a few times which will reassure him, and he will give a quiet chirrup request. She may blink some more, and will still look towards him. As he moves in behind her head he may well give another chirrup, before taking an initial hold with his mouth and stepping over her back with his far front paw. The initial neck grip may cause her to open her mouth and give a quiet 'protesting' meow, which she will probably repeat as he repositions for a better grip. Toms that have previous experience normally have a full grip within 15–16 seconds, and in longhaired cats it can take four or five mouthings to position fully. The tom's teeth do not usually penetrate the skin, for this is an inhibited grip rather than a bite. Just like that of kittens being carried by their mother (see page 60), the queen's position is dangerously close to that of prey in the tom's mouth, so it is vital that his grip is inhibited. It is thought that the queen's

posture and the tom's sexual drive produce this inhibition. Males do not act aggressively but instead solicitously towards the female. This may be due in part to her favour being of importance when in a free-living state, as well as to the need for her to maintain a lordosis position to enable him to mount.

The Mating Act

The build of the Persian (Longhair) is one that has been changed most dramatically from that of its wild ancestors, with the result that Persians today have smaller litter sizes and higher stillborn rates. Yet their mating sequence is little changed. While Persians start oestrus some months later than the average house moggie (which is at around five to nine months), in this they are not dissimilar to free-living cats. When the tom mounts the queen, he begins a treading movement with his hind legs – in modern Persian males these have been measured at a rate of around 1.6 treads a second. At the same time he arches his back, moving towards her vulva to achieve intromission. As he does so he begins pelvic thrusts. In breeding cattery conditions, full intromission to withdrawal is known usually to take less than 10 seconds.

It is possible to tell if intromission has taken place due to the unambiguously dramatic response of the queen. She initially begins by making a growling sound, then after some 6 seconds she will increase in volume and begin to twist her head to

EYE SIGNS

Throughout mating cats adopt quite intense expressions, and just before entering lordosis the queen's pupils can be seen to widen as if she is alarmed. The tom's pupils are at normal dilation during the rest period, as are hers again at lordosis. The queen's pupils widen again during intromission, withdrawal and rolling.

one side and then the other. In such a 10-second sequence of rising yowls the tom will have held the queen firmly by the scruff of the neck until almost the last second. In that final moment she will wrench free with her rotation, and then threaten to or strike at the male with the paw on the raised side of her head rotation.

Post-mating Behaviour

The male will retreat for a short distance, and as he does so may initially give an anxiety lick in response to her action. He will often display a degree of agitation by flicking his tail a couple of times. The queen then washes her urinogenital area urgently, but after fifteen or so licks begins to roll violently. In ten consecutive monitored mating sequences with the same pair of Persians, the queen rolled back and forth on her back on average a dozen times, before beginning to wash her vulval area urgently once again.

Initially, the tom watches her rolling, but then he also washes his still erect penis and surrounding fur, although not as earnestly. Both cats then lie on their stomachs with forepaws tucked under their chests, resting, ready to start the cycle again.

Left: During mating the male grips the queen firmly with a nape hold

Centre: During intromission the queen begins to cry out, building to the scream seen here at the point of withdrawal. Her mouth is open and her eyes are wide

Right: Within a split second she wrenches free from the male's hold by turning around, and swipes at him with a raised paw

Pregnancy and Birth

Behaviour During Pregnancy

You may suspect that your cat is pregnant when she lavishes more affection on you than usual. Stray (as opposed to fully feral) cats sometimes gravitate back towards people when they are well developed in pregnancy and progesterone levels make them more relaxed. However, if a female still has sub-adults from a previous litter around, she is likely to react against them. This will be more apparent in domestic-living cats who, due to indoor heating and lighting, can produce two litters a year. In feral cats this is less likely to occur as, subject to outdoor conditions, they will probably have only one litter per year, just like their wild ancestors.

As the queen goes into the last three weeks of her nine-week pregnancy, she will spend more time on her side to relieve the weight of the developing foetuses. She will also try to seek out a suitable nest site as she approaches her term. If she is a house cat, her owners may offer some deluxe bedding in a cosy cardboard box. However, she may well reject it for a more secluded spot. If this happens, do not be offended: remember that in the cat family in general strange males, not of her group, may kill her kittens. To avoid this risk (plus that of predators who would take advantage of small helpless kittens when she is away from the nest), the survival of all her ancestors depended on their choosing secure nesting places. This instinct is still strong in the queen. With feral cats, I have found queens using wiring culverts, the underneath of sheds and temporary buildings, brick piles, timber stacks, old bins and even the wheel arch of a car as nest sites.

The job of rearing the young falls entirely on the female's shoulders. However, in this connection the male is not a waste of space: he is the provider of space. By establishing his large territory which encompasses hers, he not only ensures that she has an area in which to catch food, but that it is guarded from intrusion by strange males. The male cats of the queen's group living feral may not provide food directly for the kittens, but they make it available.

Birth

As she is about to give birth, the mother-to-be washes herself thoroughly around her mammary glands and genital area, which the kittens will come into contact with on birth. She spends more time in the nest before the birth, so that the kittens will find it filled with her scent. During labour the queen may well purr, and to ease the kittens' delivery she will sit with one back leg raised. This also enables her to lick away the amniotic sac of each delivered kitten, and it is in this same position that she cleans her genital area between deliveries.

There is not a 'standard time' for delivery of a litter of kittens. While individual births can take only 15 minutes, there can be gaps of several hours between deliveries.

The young at birth are wet and messy, but the mother does not just clean them up as she washes them. She dries the hair of their coats and aligns it; by doing so she is insulating them, which is vital for their small, uncoordinated bodies. This process also allows the mother to become familiar with her new brood. (The initial unfamiliarity has allowed instances of 'fooling' the mother by planting a kitten from another litter, or even another species, in her nest.) The mother's washing also stimulates the kitten to take its first breath. She severs the umbilical cord with her teeth, and when the afterbirth is expelled she will eat it for nourishment, enabling her to stay with her kittens for the vital first few days.

Towards the end of pregnancy queens spend more time on their side or back to seek relief of the weight of the unborn kittens

Above: Severing the umbilical cord, and eating the afterbirth, after giving birth to a kitten

Top right: Washing in readiness for the next birth

Below: Giving birth to a new kitten

Above: Licking off the membranes

Below: The new-born kitten

Left: The contented family, cleaned up and feeding

Cats as Mothers

Kitten being gently lifted by its mother

Cats usually make most attentive mothers, and for the first four weeks of their kittens' lives they will spend over 90 per cent of their time in contact with at least one of their kittens. For the first four-and-a-half weeks the mother will spend up to three-quarters of her time in the nest with her kittens. From that point up until weaning at eight weeks, the mother's direct contact gradually reduces to just half of her time. A feral mother has to spend a significant part of her waking time hunting for prey to bring back to the kittens during this transition period. Before that, hunting provided just for herself and for milk manufacture.

The mother does not receive direct hunting support from the male and so has to leave the kittens in a safe lair. Rearing her young alone means her kittens are particularly focused on her actions for their survival. Consequently, if she gives an alarm growl they will instantly stop their play and dive under the nearest cover.

First-time mothers give birth to smaller litters than experienced mothers, regardless of their age. They may be less competent in their cleaning of the new-born kittens and themselves, and they may not use the proper neck hold on kittens when moving them. If the mother becomes at all anxious over the security of the nest site, she will pick up the kittens individually by the scruff and move them to a new site. The kittens remain quiet as they are moved. In

NIPPLE PREFERENCE

Unlike the mêlée in a litter of dog pups, kittens have been shown to demonstrate some nipple preference within a few days of being born. In large litters there tends to be more scrambling for place early on, but this settles down with time.

Across the eight weeks to weaning, kittens gravitate to a particular suckling position. The white cat pictured below had five kittens, and feeding positions were checked each week. Although the kittens would move over each other, there was a remarkable degree of

Suckling in cats takes up a significant part of the mother and kittens' waking day, and ensures the kittens stay quietly together

loyalty to feeding position, and it did not become more random as the weeks of suckling progressed.

These kittens showed a noticeable degree of nipple preference within hours rather than days of birth. The middle positions were locked onto by the two kittens that were primarily to hold them for the entire suckling period until weaning. Initially, the mackerel tabby that was to monopolise the teat near the mother's head for the eight weeks lay exhausted, and the two kittens that would primarily feed towards the tail position divided their time between the main positions they would later adopt and the front. Once the mackerel tabby was up and about, it took control of the head-positioned nipples.

A major advantage for cats in having position loyalty is to avoid teat laceration by sharp claws. Additionally the kittens may be displaying a form of territoriality. It is very possible that the prime position is the number one slot near the mother's head, where caring licking is also on offer, plus visual cue reinforcement from the mother's face. The farthest place from the mother's head can be the least favoured: the suckling time of the black kitten in the family was much less than that of the other kittens, and it weighed less than any of the rest.

MATING IN FREE-RANGING CATS

When a feral queen comes into heat, as her core area is within the common core area of the group, the males of her group will have territorial priority. Other males from nearby groups (and also intact house cats) may violate this territory, resulting in caterwauling. Similarly, among free-ranging house cats any intact queen wandering outside the house will attract local and more distant males, and these will run the gauntlet of intact and neutered house cat territories.

Males gather when a queen has her first oestrus, which can be as early as three-and-a-half months old, although it is normally between five and nine months. There is a tendency for the oestrus in group females to synchronise, and being around intact toms can also bring a cat into season, so these factors can increase the likelihood of mating young.

One of the few studies to look specifically at the mating of feral cats was carried out among the ancient ruins of Rome by Eugenia Natoli and Emmanuele De Vito. They found that while throughout her oestrus period a queen would have interest shown by an average of fourteen cats, at any particular session there would be around six. In their study of these free-living cats, the twenty-four queens began to come into oestrus in the second week in January and continued to do so throughout February, and the average oestrus length was nearly four-and-a-half days. A second oestrus in April took place only among those seven cats who had aborted or whose kittens had died. This second oestrus lasted just a couple of days. This is entirely consistent with my finding that Amsterdam's feral cats had one main littering

peak per year, in April and May.

One unnerving feature of cat group mating is the way in which males other than the one currently mating the queen sit quietly and patiently wait. Among the Rome feral cats, only one in three mounts achieved intromission. Intriguingly, the male that mated more than any other did not achieve a single intromission! However, when cats live together in groups almost 94 per cent of queens have been found to be able to produce a litter each year, which is a higher rate than when cats are one-to-one as in stud conditions.

An Italian feral cat diligently washes her kitten

the wild, such a move can be dangerous as the kittens are left alone at each end of the journey, while the mother makes successive trips.

The Nursing Mother

For the first three-and-a-half weeks of the kittens' lives the mother makes herself available by lying on her side, exposing her teats. With four pairs of milk-enlarged mammary glands, this is also a position that gives her some relief. Up to six weeks of age the kittens' demands increase, but although the mother is still available she becomes less willing to provide them with milk.

In litters up to the number of teat pairs, that is up to four kittens, the new-born are usually of comparable weight. With litter sizes from five upwards the average kitten birth weight is lower. Although with larger litters the mother makes more milk, the supply is not infinite, so the average weight of kittens from larger litters is not as high as that of kittens from smaller ones.

Once the kittens have reached a certain weight, the reality of milk-flow restriction forces a change in the mother's behaviour. The larger the litter, the earlier she begins to cut down on the kittens' accessibility to suckling, either by moving or by lying on her nipples.

Stages of Kitten Growth: 1

New-born

When kittens are born their eyes are tight shut, yet they find their way to their mother's teats to suckle. For the first few days their responsiveness to sound is minimal. However, when the unborn kittens are only four weeks on from conception they have already started to develop their sense of touch. Similarly, when born they have a reasonable sense of smell, although this will not be fully functional until the kittens are three weeks old. Their sense of temperature recognition enables them to home in on warmth from their mother and litter mates, and to retreat from the cold. The combination of touch, smell and warmth allows the kittens to find the rewarding source of food. For the first week or so the kittens also have a 'rooting reflex', whereby they automatically push their heads into warm places.

New-born kittens do not have the muscular strength to support their bodies off the ground at all. Despite being fur covered, they are weak and vulnerable at this stage. Not only are their eyes tight shut, but their small round heads lie flat on the floor on their chins virtually all of the time when they are not suckling. They exhaust quickly and it is a major effort for them to haul themselves blindly up onto a teat. They also pull themselves to their mother's head and will push against her warm, moist mouth. The mother nuzzles them gently when necessary.

One Week Old

By the time the kittens are one week old, their prolonged feeding will have had a noticeable effect. Their bodies are much fuller and more rounded as they have been gaining weight rapidly. Although their tummies are still on the ground, their limbs have more strength and they can keep their heads up, but they do spend a lot of time sleeping – when they aren't feeding!

As the kittens' figures have filled out, so have their heads, although at this stage their ears are still very small. Their eyes are no longer sealed as tightly, and a gap of eye can often be seen in the corners nearest the nose. In consequence, their appreciation of what is bright and what is dark should be improving.

When the mother moves away from the kittens to feed or rest, they usually remain huddled together, like a pile of sausages. If one does move out, it soon moves back towards the group, but only as fast as its extremely shaky limbs will allow, with its body rocking back and forth as it moves slowly along.

New-born kitten   One week old   Two weeks old

Two Weeks Old

The mother cat spends a considerable amount of time washing and grooming her kittens. Part of this is to cause them to roll over or get up and so allow the mother to stimulate them to eliminate, so that she can lick away droppings that would otherwise foul the nest. The kittens' eyes are now open but they are very cloudy, with little ability to bring the world into proper focus. When not with their mother, the kittens still prefer to spend time huddled as a warm group, lying in close contact with each other.

The kittens' progress is still on their tummies with wobbly legs pushing them along, but a week makes quite a difference. They move more readily across each other and are more venturesome, moving with heads bobbing but still held very low. They continue to gain weight rapidly, but although their ears are growing they are still buckled down. It is also at this stage that some kittens begin to place their paws beside their mother's teats, but as yet they do not have the power to paddle.

Three Weeks Old

The kittens are beginning to look cute now, rather than just like blobs. Their ears are firmer and more upright, but still with a bit of a downcurve. They have begun to discover walking, rather than pulling themselves about on their tummies, although their tummies do not leave the ground by much. The experience of walking is still new to them and their paw control is very shaky. Their heads, bodies and tails are still held low, with their back legs pushing along. However, they will now look directly at you, although they still stay close to mother. As the kittens are filling out in size, the point has come when suckling becomes a tight fit!

Four Weeks Old

The kittens' ears are now properly upright. They can walk about with their tummies well off the ground, although they are not yet fully upright. However, they can and do stand tall on their legs experimentally sometimes. They now often hold their tails upright, where previously this had not really been possible. With their greater mobility, the kittens are becoming much more adventurous. Their eyes have become clearer, which helps their investigations; it is also an advantage for their increasing play with litter mates.

PAW PADDLING

From around three weeks of age, the kittens will stimulate their mother's milk flow by paddling at her with their front feet. These kneading movements are exactly the same as those made by the contented adult cat as it settles down on a warm, soft surface, such as its owner's lap.

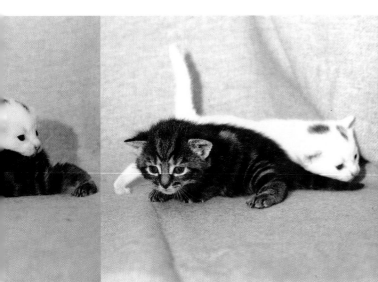

Three weeks old

Four weeks old

Stages of Kitten Growth: 2

Five Weeks Old

At this age the kittens suddenly become much more lively and mobile. They will focus on and pursue play objects and will even clamber up things, including people if they are sitting down. Their eyes look crystal clear, and their hearing is now fully developed.

This increase in the function of their senses has coincided with the kittens' increased ability to move about. They seem to be always on the go, but then they will suddenly nose dive and take a nap. They try out their teeth by gnawing on baskets and just about anything else, and have certainly developed the cats' curiosity to find out what is behind things. Their play is becoming more vigorous, with batting occurring. At the same time, their mother is cutting down on stimulating elimination and licking away the excreta. This was an important survival ploy when the kittens were less mobile, to prevent a detectable mess accumulating around the nest.

Six Weeks Old

The kittens have now undeniably entered their cutest phase, with perked ears and adorable faces. Their senses are really attuned and they are now walking about fully upright on their legs. The amount of time they spend playing with objects really increases. Roll a ping-pong ball gently towards them and they will be keen to investigate and bat it. If they are living with a mother that is bringing in prey from the outside world, they will be keen to grip it and will growl in possession. If you play with toys of similar size or with a cord, they will grasp that in their mouths and will show a determination to pull! They can be seen playing gently with their mother's tail, or with their own or those of their litter mates.

By this age the kittens are becoming more self-contained, in as much as they will not only initiate play with a dangling string, but will also sit quietly between some activities. Climbing about has become one of their major preoccupations! As part of the mother cat cutting down on suckling during weaning, she sometimes sits upright, but that doesn't always prevent an enterprising kitten pushing in.

Post-weaned Kittens

Weaning is the transition from milk to solid food for the developing young and in cats is usually accomplished by the time they are seven weeks old. As the kittens have been growing, so have their food requirements, in part at the expense of the mother, who during the lactation period will have been losing nearly 6g (¼oz) a day in weight. However,

Five weeks old   Six weeks old

KITTEN ABUSE

Mother cats are renowned for their caring devotion to their kittens. Sometimes things can go awry, but fortunately it is unusual. An example of this occurred in a young family that I monitored.

The mother cat showed due attentiveness to her kittens, fed and washed them all appropriately, and attended to them if they called out during the early weeks of kittenhood. It was a hot summer, and as they grew towards weaning she would lie away from them to cool off. However, at this stage she began to wash the smallest black kitten over-zealously, and her attentions became rougher. She would hold him around the neck with her paws, biting and kicking him – actions sufficiently violent to cause the kitten to cry out. These actions were in complete contrast to those she displayed towards all her other kittens, although she did continue to wash and suckle the black kitten along with the rest of the litter.

There was, however, a behavioural price to be paid for the mother's actions. By the time it was weaning, yet still taking milk from its mother, this kitten was noticeably more fearful than the others. Although all the kittens had been handled regularly since they were very young, a finger proffered towards the black kitten provoked a defensive hiss. In contrast, its slightly larger brothers and sisters would, by this stage, move forward to investigate the finger.

At this point, a transformation occurred in the relationship between this kitten and some of his litter mates. While play between all the kittens was becoming naturally more vigorous, the fearful black kitten began to be attacked more determinedly and for prolonged periods by other kittens. Its mother's actions seemed to have made it more fearful, which in turn made it a 'victim' for its litter mates. Such early conditioned behaviour is usually hard to break in later life.

despite any discomfort their sharp teeth may inflict, the kittens will continue to try to suckle.

By eight weeks old, the responses of the kittens are becoming more mature. By the time they are ten weeks old, although they are still smaller and slimmer, their proportions and posture are virtually those of an adult. The kittens' motor co-ordination is now well developed, and they would have the movement control of an adult if it were not for their lack of experience and youthful exuberance. Ten-week-old kittens rush about in sudden spurts of enthusiasm, chasing around after their litter mates. They stalk earnestly and move with sudden skittishness, seek out cover and pounce wildly. However, from nine weeks on their play becomes more earnest, until around fourteen weeks of age the aggressive edge dulls litter-hood ties. The kittens are now ready to become independent juvenile cats.

Seven weeks old

Eight weeks old

7 Play and Learning
The Role of Play

'When I play with my cat, how can one know if I am entertaining her, or she me!'

Montaigne, sixteenth-century French essayist

We recognise play when we see it, and that it is more common in juvenile cats than in adults. It seems to be practice for life-skills development without the dangers of such adult characteristics as aggression, yet play has been harder for researchers to evaluate than other behaviours like predation. That in itself is partly a comment on our relative perspective, but it is also true that predation is less ambiguous than play. Much of what is commonly termed 'playing with prey' is not play but the inhibited form of dispatching prey (see Chapter 5). However, when such sequences are more prolonged in house cats than in hunting feral cats, it can seem as if playfulness has been incorporated. When hunting moves are made by a cat against inanimate objects, such as a ball of paper, the component of play is easier to identify. This is particularly so if the cat initiates the event.

When adult cats play with us, it is usually considered to be a retention of juvenile character due to our size and support. One factor that is usually overlooked is that our care of house cats gives them significantly more free time. They do not have to catch prey to survive, nor do they have to patrol huge territories. The drives for these behaviours are nonetheless present and will surface as play.

Although adult cats play less than kittens, they will play by themselves, not always requiring our presence for stimulation. To compare kitten rates of play with that of solitary house cats is not a fair comparison, for a good proportion of kitten chase and pounce moves involve the other kittens in the litter. In that sense, a fairer comparison would be with adult cats playing with us.

Play is most recognisable in species like the cat, whose range of behaviour is not fully present from near birth and who have complex nervous systems. Play has been called paradoxical behaviour, in that it often causes the reverse of the behaviour sequence it mirrors. For example, a successful adult territorial fight should cause the opponent to withdraw, while kitten fighting play is part of the socialising between litter mates.

Cats brought up together from kittens together behave as if they are litter mates, and due to their relationship with us live in an extended kittenhood (neotenous behaviour)

Oriental tabby behaving playfully as if patting a small mammal prey

PLAYING WITH TOYS

When we think of play, we automatically think of toys. This is very much a human response; nevertheless, as much of kitten – and adult cat – play involves moves used in hunting, 'toys' often take the place of the prey. Such toys need not be expensive shop-bought items: anything small that will move easily can play the part. This includes tightly rolled balls of paper and, of course, the traditional ball of wool. Manufactured toys are often too heavy and hard to make good playthings for cats and kittens – good 'toys' should be light, so that they move a long way for little effort.

Learning Through Play

Play Development

Play starts in kittens only when it is physically possible. In their first two weeks of life the kittens mainly lie huddled together with minimal mobility. By three weeks they have more movement but are still shaky. At four weeks old they have become mobile, but footwork is still hesitant and they clamber over litter mates in a clumsy way, still without sufficient co-ordination and flexibility for properly engaged play. By five weeks old things have changed and the kittens' mobility and play have increased together, and by six weeks they are leaping, climbing and full of a playful zest for movement. They now have good paw-to-mouth co-ordination, which enables them to play with and grip dangling string.

The early cumbersome movements develop into play fighting, then as weaning approaches play hunting moves are seen – classified in one system as 'mouse pounces', 'bird swats' and 'fish scoops'. A problem connecting tidily adult behaviours and those of kittenhood play is that more than one piece of behaviour can relate to a kitten movement. The 'bird swat', for example, is used extensively in the recapturing of birds (see Chapter 5), and is also that of standing defence (see Chapter 8). Similarly, the 'fish scoop', which is not as frequent a play move as the bird swat (but is particularly feline), allows the cat to reach into holes and spaces and root around for prey.

An alternative classification (which does not make an interpretation of adult use) describes kitten behaviour when playing with objects in such terms as 'scoop', 'grasp', 'poke' and 'mouth'. When kittens interact socially with each other their positions have been classified descriptively as 'chase', 'vertical stance', 'belly up', 'pounce' and so on. Yet each of the above moves used on objects is also used in social moves with other kittens. One kitten will bat another or make a scoop move at a tail. Similarly, social moves like pounce and chase can be used on a small ball.

The amount of object play that kittens engage with is more than by less obligate carnivores, so a fair part of object play is linked to predatory behaviour. John Bradshaw has made the astute observation that although this object play has been compared to hunting, it is most comparable to post-capture 'play'. In the rough and tumble of play, if one kitten is standing up on its back legs, the responding kitten will go down on its back, and vice versa, in moves similar to those of a fight.

One feature that does change with growing mobility is the increasing distances covered as post-weaned kittens race about in their games. This frolicking and running about with mad dashes for what looks like the sheer fun of it, is called superfluous play.

Play Signals

The purpose of play is play. In kittenhood, a play attack is usually just that, and even when ears can be positioned to express threat or defence (see Chapter 8), they are not usually so positioned. I believe kittens recognise that a threat is not real, and so a real adult response need not be made. Part of the reason, however, is physical: during early development kittens cannot make such signals due to the immobility of their ears.

- New-born to one week old – ears are round and immobile.
- Two weeks old – ears are developing but bent over all the time and still immobile, and in consequence look permanently defensive.
- Three weeks old – ears have become erect, but are fat and devoid of much movement.
- Four weeks old – ears are fully upright.
- Five weeks old – ears are increasingly flexible, allowing them to flatten but still not to fold properly at the back.
- Six weeks old – ears are big and can fold on contacting something.
- Seven weeks old – ears can now fold properly, but in the mayhem of continuous play kittens do not normally give aggressive ear signals.

Play, however, can become reality, as it did for the small black kitten of the litter mentioned on page 65. When it was bitten in earnest by a litter mate, causing it to cry out, it suddenly put on the full defensive ear position as it was genuinely fearful.

TAIL PLAY

At seven weeks of age tail play is a real pull on kitten attention! Typically, a kitten will begin patting movements, following the tail of a litter mate. The litter mate, though, may be busy with some other play action and not react. After a few seconds of tail following, the first kitten's attention is usually drawn by something else. It may be that tail following is hierarchical, for some kittens seem to be more tail-followers, while others are more tail-followed.

In older adult house cats, tail play is often quoted by owners when talking of their cats 'having a sense of humour'. The cat will set itself up for a period of self-motivated play by getting up onto a sofa back or staircase post and curving itself around, with its head down and foreleg extended downwards. Once in position, it will friskily attempt to grab hold of its own tail. This often takes place when the cat is in a heightened mood, perhaps because the owner has returned home. Owners will often collaboratively play with their cat in this heightened state, by touching its back or putting something where the cat can grab it.

Kittens' moves are signalled as play by their exaggeration and lack of aggressive intent or intensity, which means bites are inhibited. Although litter-mate play around weaning usually lacks the aggression that can appear in older kitten play, there are weight and confidence differences between kittens. Within the close quarters of the nest area a litter hierarchy develops. The nervous behaviour of underweight kittens can evoke more forthright responses from heavier kittens.

The development from four to six weeks sees a remarkable transition in the kittens of mobility and consequently in play, involving litter mates and also objects that can be investigated as if they were potential prey

Learning About Prey

Sensitive Periods

Learning is in part about getting ready for adult life. As they go through their various stages of development, kittens progress through what are referred to as 'sensitive periods' (previously called 'critical periods'). One such period involves learning the skills used with prey, and another learning about litter mates. While cats can still learn skills later in life, there is a correct window in which they are programmed to assimilate information at the appropriate stage of their development.

Around four to five weeks old kittens undergo a major transformation. If she has access to the outside world, the mother cat will start to bring in dead prey for the kittens in anticipation of weaning. Up to this point, in the isolated security of the mother's nest area they are unlikely to have met other animals, dead or alive, or any food source other than milk from their mother.

When the mother initially begins to bring prey to the nest she will growl and keep the prey. Her introduction of the kittens to prey and prey-catching is not initially as an instructor. Instead, she portrays herself as an apparent competitor, which stimulates the kittens' attention. However, she avoids frightening them by intermingling her growls with encouraging purrs.

Observations I made recently confirms that the first meeting of a kitten with dead prey brings about a transformation in behaviour. From mild interest and kittenish reactions, the four-and-a-half week old kitten suddenly became alert and focused when the dead prey was lightly tugged in front of it. It held it with its jaws and began growling repeatedly in a possessive manner.

The kitten then sniffed its prey – a dead mouse – repeatedly for 18 seconds, and did not follow the mouse when it was pulled away, but sniffed the area where it had been. After a further 9 seconds it began to mouth the prey. But when the prey was repeatedly pulled away, as if a competing kitten had got hold of it, the kitten's response became more serious. It sniffed the prey repeatedly for a further 9 seconds, and when the prey was tugged away again it restrained it with

its paw and made its first bite, a correctly orientated nape bite, a mere 11 seconds after its initial sniff. The kitten's whiskers were around the prey, and it pulled at it before it was pulled away again. Again the kitten tried to restrain it with its paw, but again it lost the mouse, causing the kitten to give a quick nose lick. The prey was dragged along and the kitten followed it by smell before trapping it again; it breathed in sharply, then gave five deep breaths before biting the haunch of the prey while putting its paw on it. After adjusting its hold it shook the prey vigorously from side to side, lifting it off the ground. Still holding it firmly by the rump it finally dragged it away from the 'competitor'.

Kittens need to learn that food is contained within dead prey. With soft-bodied, small prey such as voles or mice, play and seizure often puncture the skin and so a connection to food is made. With stiffer-furred or feathered animals, the physical barrier can inhibit even some experienced hunters in making the connection. Consequently, captured wrens are more likely to be devoured than blackbirds. The introduction to prey in the rough and tumble of the kittens' nest makes it more likely that they will make the connection.

THE ROLE MODEL

By the time kittens are nine weeks old their mother is still taking prey back to the nest, but by this stage it may still be alive (although she may kill it before a kitten takes possession). The similarities between this and our house cats taking prey back to their home and releasing dead or live prey into the kitchen (if that is where they are normally fed) is remarkable. By working through a gradation from bringing home initially only dead animals progressing to increasingly lively prey, the kittens become more expert at dealing with it. While the mother is bringing back food primarily for consumption, she is also able to display the technique of killing prey. However, some isolated kittens manage to kill without that input, showing that it is not essential to learn it all from their mother. Nevertheless, kittens that are able to watch their mothers kill are more successful in hunting.

Learning About People

How Cats Relate to Us

When the mother begins to bring in prey for the kittens as they approach weaning, and they learn how to treat it as prey, they are relating to another animal in a distinctly different way from that in which they relate to their mother or litter mates. Yet if other species (including those they have come to recognise as prey) were introduced earlier into the nest area and were tolerated by the mother cat, then the kittens would behave towards them in an equally tolerant way.

Once young kittens' eyes begin opening and they start to move about, they begin to encounter other animals around the nest area that have been tolerated by the mother. These are accepted as litter mates, for normally the mother would fiercely exclude everything else. Whatever they are, these tolerated animals will be accepted by the cat even in adult life.

If we wish our cats to behave towards us as if we were litter mates rather than the enormous alien species we are, then it is vital that kittens are habituated to people in that window of time available for recognising litter mates. When kittens are brought up with non-cat species, such as puppies or rats, the kittens show a similar distress when those animals are taken away to when a feline litter mate is removed.

Eileen Karsh established that the window of the sensitive period for habituating kittens to people is from two weeks to seven weeks. Cats handled regularly at this time were prepared when older to stay on someone's lap longer and approached people more readily. Cats that were handled after this period, that is from weaning onwards, behaved no differently to cats that had not been handled at all.

Unfortunately, this corresponds with the very time, post weaning, when kittens are normally taken to their new homes from catteries or rescue centres. All too often, as young kittens they will not have been handled regularly. Brought into new homes where they are the focus of attention they receive a lot of handling, yet it seems this will not socialise them, unlike a little time spent each day before weaning. Consequently, as adult cats they

Playing gently with young kittens before they are weaned helps them relate better towards people for the rest of their lives

will not relate to people as well as they might otherwise have done.

The problem is that it would be traumatic to remove kittens before weaning to their new homes so that their new owners could relate to them, for this deprives them of their mother's attention and makes them overly dependent. However, Karsh found that it was not necessary for the habituators to be the eventual owners; they just need to be people. On the other hand, kittens do bond to particular people, so perhaps the ideal is to spend some time, regularly, well before weaning at the cattery with the kitten you have chosen. This should be in the nest area in the presence of the mother, who must be trusting.

Timid and Dependent Cats

Eileen Karsh's study found that around one fifth of cats did not respond to handling with any improvement of socialising due to their timidity. In adult cats there are two clear patterns of nervous, unsettled behaviour – the timid cats and the dependent cats. However, this distinction is only noticeable in connection with their owners. Without the owner present, the dependent cat's wariness may seem little different to a completely timid cat who is not socialised to anyone. In reality, though, such a distinction is rarely of practical value, as most anxious, dependent cats can be quite timid when they are not gaining confidence from the presence of their owner.

Timid cats are vulnerable to stress, and some of their characteristics are induced by that stress. They can be generally very wary of people, disturbance and noise, and will seek out refuges and avoid social contact. They do not readily explore if placed in new surroundings, and will often sit still, stressed and nervous. They generally run sooner from strangers than do more confident cats. Karsh's findings suggest that there may be a genetic predisposition to timidity in some cats, but that handling early in their lives produces better socialisation in most cats, and as a consequence less timidity in adulthood.

To find a measure of timidity, and how people affect such cats, Karsh timed how long cats took to come out of a compartment to investigate. When no one was about, confident cats wandered out within 18 seconds, but if a person was there they came right out in 3 seconds. In contrast, timid cats took 86 seconds with no one about, and nearly the same, at 75 seconds, if someone was there.

I found the effect of not only the lack of early kittenhood habituation but also the lack of adult human social interaction to be quite dramatic when free-ranging feral cats were taken into captivity and placed in a confined space the size of a small room. When a person walked nearby, some of the cats crammed themselves into the narrow gap behind

Continuing habituation with older kittens

the door. These were timid cats, in the sense of not having been habituated to people, but in their normal activities were competent and capable.

Similarly, among house cats there are many territorially confident males, with larger ranges than the local average, who are avid, successful hunters and trusting of their owners, but who rapidly 'evaporate' if strangers are about. There is therefore a distinction between timidity in response to people from lack of habituation during kittenhood and later life, and general timidity.

Coping with Ambivalent Cats

Nervous, dependent cats are the 'one-man dogs' of the cat world, keeping a dependent fixation upon their owners, yet remaining very shy of other people in general and exhibiting 'nervy' behaviour. Within a family they can gradually broaden their acceptance of close people.

Kittens stand a higher chance of being nervous, dependent cats if they were small kittens relative to the litter average, were taken from their mother too early before being fully weaned, and were not properly habituated during development to a number of people. Kittens handled by more than one person are more accepting of strangers in later life than those handled by only one person.

Nervous, dependent female cats can exhibit an apparently strange piece of behaviour, even in response to their owners, when they are flustered and anxious. A hand run down their backs in a normal gentle stroking manner can cause them to lower their backs as they walk. If the hand is lowered, they further lower their hindquarters in the manner of a limbo-dancer. However, once such an animal is reassured, her anxiety will instantly melt and she will not only stand to be stroked but will raise her rump firmly and rub around her owner's hand. Such cats have been parodied as a 'neurotic mess', flipping back and forth between nervous retreat and anxious clinging. They produce this and other ambivalent behaviours due to their conflict between wanting to make contact and at the same time fearing it. They may run to the cat flap as if fearful, but return as fast if called, then retreat at any movement, only to return once more when their owner calls again in exasperation.

Nervous, dependent cats will sit on your lap more frequently, spend more time in your

Above: Habituating kittens is a rewarding task

presence, knead you more with chest paddling, and rub near you or on nearby objects more frequently than most cats. Due to the confidence gained by your arrival they will want to rub, but because of their anxious lack of confidence they will mainly rub on objects near you. These moves are the equivalent of 'air-kisses' in people!

The nervous, dependent cat is also prone to behaving in a depressed manner. When put in a cattery, it may refuse to eat and so lose weight, for in this situation it is deprived of you, its focus, as well as its home.

Left: A nervous, dependent cat rubbing against its owner's legs

Let Your Cat Have Fun!

Adult Learning

It might be assumed that as early conditioning is so important to cats, learning in adult life is insignificant. However, while it may not have the same ability to fix the pattern, later learning is nonetheless significant.

Owners of nervous, dependent solitary cats can often lead quiet, enclosed lives. This is because these owners, knowing of their cat's timidity towards other people, strive to limit their cat's stress by limiting their own social lives. This may reassure their cat, but it is not ideal. One solution that would not be anticipated by such owners, but which I have known to work, is to go away for a couple of weeks and have a group of noisy young people house-sit! On your return you may be amazed to find how your cat has habituated to a wider circle of people and is less stressed by a stranger's presence. You may find it less of a 'diving in at the deep end' exercise to introduce other people in stages. Your acceptance of people will help your cat to become more tolerant, and gradually habituate to them.

Similarly, although missing the window of the hunting-sensitive period of kittenhood reduces the effectiveness of the cat as an adult hunter, bringing prey back into the core area of its range enables some adult education catching-up to occur. With the catch and recapture of 'playing with prey' taking place in front of other group members, the cat has an opportunity to observe and take part in the process by relieving the hunter of its prey or catching it when it escapes from the hunter.

This has happened with my own cats. Leroy is the ultimate competent hunter, while Tabitha was originally hopeless. However, while still not anything like as capable as he is, by observing and taking part over the years she has become reasonably competent, particularly with 'easy' prey such as shrews.

When a bird flies off or a mouse escapes, it is usual for the hunter cat to retrace some of the

In the cat's 'teenage' months hunting a 'cat wand' toy substitutes for the real thing

scenes of action of the hunt. Where other cats have been watching, inhibited from taking part by lack of territorial confidence, after the escape they may retrace the scene by sense of smell. Such events advertise to other cats that 'this is a good place to be', and provide not just a demonstration of skills but vary the contact with prey.

Playing with Your Cat

Adult cats often like to play grabbing and holding games with owners, which the cat will often initiate. Staircases provide great places for cats to play. When the cat is on the stairs and a pencil or something similar is moved along past the other side of the stair-rails, in and out of sight, it allows the cat to carry out a number of the moves it normally uses in the capture and recapture of small mammals. The cat will not only thrust a paw forward and grab the pencil, but will have to anticipate, select and change the gap through which it thrusts its paw. Many owners enjoy playing with their cats like this, but as the cat is employing a hunting grab it will be using its claws for the strike, so take care! The cat will also try to grip the pencil and take it to its mouth to chew.

Most cats enjoy pursuit games – anticipating, pouncing and running after a string, or the modern version of a 'cat wand', which is a flexible rod with a string and small toy on the end. For kittens around the 'teenage' time of three months old, or the thin-bodied form of Siamese that don't slow down much in adult life, the wand gives the cat the fun of pursuit without perspiration for the owner!

More confident cats will often initiate pouncing entrapment games through material with their owner, where the cat can see and respond to the movement alone. Owners in bed often move their toes around under the covers to encourage their

CAT GYMS

The advent of the cat gym has been a real boon for many cat-owning households. I was initially sceptical of their value, thinking that they would be just an encumbrance in most homes. They are, after all, only carpet-covered platforms on posts. Indeed, they can be white elephants, unused and gathering dust if sited wrongly and their use is not encouraged by owners. One of their greatest values is that they provide a place that both owner and cat recognise as a site for play, a place of communication. If you play regularly with your cat with the gym, then the cat will go there to request play.

These gyms also provide scratching posts, and most importantly somewhere to clamber up and from which to sit and watch the world. The gyms can be played with most joyfully by a pair of juvenile cats that are still kittenish enough to maximise play, but big enough to have adult co-ordination.

Adult house cats will play by themselves, suddenly having a mad chase after their own tails interspersed with exaggerated licks, or wild dashes around the house, or play hunting by batting around a piece of paper. As they live with people, this may be behaviour retained from kittenhood. However, much of our own behaviour is not dissimilar!

cat to pounce. Similarly, some cats will tunnel under an overhanging bedspread or discarded jumper, or go behind a curtain, and follow and capture a pencil or similar object that the owner runs across the cloth.

Far left: When interested, cats will chase after and catch gently thrown objects like cones

Left: Playing with your cat not only provides him with both mental and physical stimulation – particularly important for cats which are confined to the house – but also allows you to observe those subtleties of behaviour which you may otherwise miss

8 Affection and Aggression

Carnivores such as dogs, who hunt in packs and are less individually territorial, work within a hierarchy. This ranking has to be decided before the hunt, and so numerous aggressive interactions take place. In pack animals, the potential for aggression leading to dangerous, even fatal, confrontation is high, and as a result dogs possess a repertoire of appeasement behaviours, similar to those of wolves and hunting dogs. In contrast, cats are intensely territorial, and consequently the ways in which they deal with social interaction and aggression are very different.

Social Bonding or Appeasement

Historically, before domestication solitary hunting cats could not reach high densities, so there was little need for appeasement behaviours. Even today, cats prefer to threaten, fight, withdraw or avoid. This relative lack of appeasement moves may be what makes overcrowding particularly stressful to cats.

When domestic cat groups come together it is for social activities rather than for communal hunting. When they are held together by the accessibility of food, tolerance rather than co-operation is needed. The successful survival of the wolf depends on the pack, where aggression is necessary to establish rank, so appeasement balances it. As the cat's survival is not dependent on a pack, there is no need to establish rank and so aggression in a group is not so critical; consequently, neither is appeasement. However, the group does work by the social bonding glue of rubbing, mutual grooming and distance tolerance, which as owners we equate with affection. Yet the strategy cats employ of being territorial, and delineating territory by a system of scent and other marks, is a form of sabre-rattling in absence. Declaring presence in this way is certainly not appeasement but it does avoid aggressive interactions, unless the intruder ignores them.

GOOD FRIENDS

Cats do not normally sit close together unless they are group or family members.

In households with more than one cat, the dynamics of the social relationships between the cats can be up and down. A measure of how good the relationship is at any time is how close and how frequently your cats are prepared to curl up and snooze together. Yawning and blinking are also bonding signals (see Chapter 9).

Confined cats will still maintain their distances from each other if they do not believe they are a group.

Tail Raising

When is a gesture one of social bonding and when a matter of appeasement? In essence, appeasement aims to prevent an act of aggression and is carried out by an animal of lesser rank. As hierarchy is not reinforced by co-operative hunting, rank for cats is not particularly clear cut, so close interactive moves are more likely to be social.

Tail raising may be a request to allow contact rubbing, and such mutual co-operation is motivated by the need to bond rather than to appease. In contrast to the way in which an appeasing dog would behave towards a threatening one, tail raising and rubbing are not carried out in the face of aggression. The cat does not have a submissive vocabulary to cope with that. It is possible to recognise an 'acknowledgement hierarchy', as the affectionate move of rubbing may be initiated more frequently by a 'low ranking' cat towards a 'high ranking' one than the other way round. I believe that tail-up is not just a permission to rub, but does constitute a greeting, allowing cats to come closer amicably.

Above: One of the author's cats greeting him by tail-raising and head-rubbing against his hand (as if to another cat's head)

Below: Where contact occurred between observed free-ranging farm cats 93 per cent were bonding moves (mutual licking and rubbing), while only 7 per cent were aggressive. In farm and feral cats aggression between group cats is mild, and only serious towards intruding males. In suburban house cats and confined cats aggression is more frequent due to density itself, and when density closeness has not allowed time or inclination for contact permission signals

Body Language

When a cat's tail fluffs up in a dramatic way, its back arched in a parody of a witch's cat, it may be defensive or aggressive. Cats involved in a stand-off will be making 'mrrow' threatening sounds, but to read what is happening look at their ears. If they are down flat against the head, that is the defender. If they are down flat but with a twist so that the tips of the back of the ear can be seen from the front, then that is the aggressor. These positions can seem very similar to us, but there is no such ambiguity for the cats.

If a fight develops, the defender's ears remain

flat throughout, but the aggressor's get folded flat with a flick at the instant it dives onto the other cat. If it remains the clear aggressor, as it disengages the pinnae will lift quickly back into the aggressive position. However, if it is not intending to continue the aggression, the ears return to neutral. These ear position flips happen so fast that it is almost impossible to register them clearly at normal speed.

Look also at the cats' eyes. As a measure of its stress, the defending cat's pupils will be wide open with fear, while the aggressor's will be narrowed tight. At such times, this is part of the 'fight' or 'flight' reaction of the cat's sympathetic autonomic nervous system, which also causes the hair to stand on end. The aggressor threatens

It may look comical, but intruders beware for this cat's ears have folded so that their backs can be seen from the front, and its pupils have narrowed, clear threats of attack

Facial Expressions

Cats' faces reflect their mood, but although they indicate this to other cats, the changes are primarily to give a functional advantage to the cat. When the cats' eyes widen they gain wider peripheral vision, an advantage when anticipating being attacked. When the eyes narrow they gain better depth perspective, an advantage in judging where to attack.

A young cat in classic defensive posture: ears flat, pupils open wide, hissing, with head drawn back

with its posture by approaching sideways-on to look larger, and its head turns as it gets ready to throw itself onto the other cat. The frightened defending cat will either crouch down, prepared to strike, or go down on its back, ready to use its back legs to rake the underside of the aggressor as it attacks.

Territorial confidence will usually give one cat the advantage over another, and there will be an aggressor and a defender. However, two large, equally confident males both flagging aggression is the sign for a relatively rare, but very serious dispute (see page 84).

This is the snarl of a confident adult cat. Its pupils have narrowed with aggressive intent, but it does not feel the situation warrants any ear flagging of intention to attack

This lick is a precise move in which the tongue goes up to cover the cat's nostrils and assist the Jacobsen's organ. It occurs in a number of situations, notably flehmen (see Chapter 1), displacement action and during aggressive interactions. Owners are usually unaware of the length of tongue extended or the precision of its movement

Defensive Behaviour

Lost Confidence

Territorial disputes can have a devastating effect upon a cat's confidence for a prolonged period. An example that I monitored over some months involved a black-and-white longhaired, neutered tom cat. Typical of so many cats, this pet lived an urban life in late Victorian terraced housing in London – high-density living, with the houses huddled together with small gardens.

The cat led a very contented life with access to its family via a cat flap, and with well used sunny and shady snoozing spots in the garden. Then suddenly the cat from hell arrived. The new male made life a misery for the resident, whose confidence was shattered. The new cat not only

Mild aggressive mealtime interaction between feral cats. One cat rears up in response to the others' incursion

went in through the resident's cat flap and ate its food, but attacked it inside its own house. Over the following six months the resident cat hardly dared venture outside. During that time, when the first cat's owners were at work the intruding cat used the resident's garden snoozing spots.

Outside territorial events can also affect the relationships within a multi-cat household. My own neuter tom Leroy was housebound, recovering from an abscess following an attack by a new neighbouring tom. During this period, and for a while afterwards, his confidence was not what it had been, and my other cat Tabitha took advantage of the situation and asserted herself, initiating mild aggression.

Other Defensive Behaviour

Low-speed inhibited fights between household cats may break out around meal times, or seasonally with winter 'trapped-in' blues. When an aggressor approaches an upright cat, the defensive cat may crouch and flatten its ears to minimise the target, but raise a wary paw and even strike the air with a mixture of hissing and growling. However, this is not a submissive pose, for the cat is ready to retaliate. If one cat runs in fast to attack a standing cat without warning (which can happen

Crouching with head drawn back and ears defensively flattened against an inhibited aggressive threat between household cats

with displaced aggression, as when a specific 'target' is missed but the steamed-up cat runs on to another, or in some territorial attacks), then the attacked cat rears up with the momentum and bats back. Both cats will stand, reared up, but the defender will use alternate front paw swipes, standing tall in the same position it would adopt when lying on its back during a full-blown scrap (see page 85).

A Cornered Cat

If a cat is cornered aggressively or is fearful that it cannot escape, with no submissive gestures to call on, it cannot diffuse the situation and thus will be dangerous, as it will use last-ditch fighting. It is this, plus an impressive armoury of fighting weapons, that will make even the largest dog which has cornered a fleeing cat suddenly stop and retreat. Dogs are more respectful of cats than they may at first seem, for there are more 'corners'

about than we normally recognise. If, for example, a dog chases a cat up a tree, the cat is secure and merely waits for the dog to go before coming down. If the same dog startles the same cat up onto a garden fence where there is no additional height for the cat to climb, the cat will feel cornered. Consequently, if the dog puts its head up or leaps up, it receives a faceful of claws as the cat strikes at it. Exit dog, fast.

It is such situations that lead to the near legendary tales of 'cat bravery' on behalf of their owners, although the frequency of such stories is far lower than with dogs. One involved a women opening her front door to find a stranger swinging a cleaver down at her head. Her cat saved her life by jumping onto his face from the stairs and scratching him. This apparently gallant act was probably due to the cat being startled, feeling trapped and cornered, and seeking an escape by clambering up!

TAIL SIGNALS

■ The cat's tail has great flexibility which enables it to provide balance, not only in fast cornering moves, and climbing, but for position compensation such as when a cat is eliminating. Its moods are also expressively conveyed by tail positions and movements.

■ Upright, with tip curved over – the cat in 'neutral', with the tail tip naturally turning back and forth in time with the walk.

■ Erect for its full length – the greeting to another cat or ourselves, often followed by rubbing. When kittens greet their mother, particularly when she has prey, they will raise their tails and rub along her in request (during earlier kittenhood she cleaned under their tails). We also witness this as a request for food. When we rub along our cat's back the tail will flick up into this position. If stroking is continued along the back, some cats will fluff up the lower third of their tail, and curve the rest as they paddle and purr. During such protracted greetings, the involvement of anal scent glands can be detected. When the cat is lying out full length with tail erect it is trying to cool down.

■ Erect for full length, quivering – the position adopted for spraying, with the cat's hind legs standing tall. This also occurs with 'air spraying', whereby neutered cats in our presence make the tail and body movements of spraying without actually doing so, when they go into an area where they are not fully confident.

■ Held to one side – during the female sexual position (lordosis), to allow mounting.

■ Tail flicks of seated cat – used to check that nothing is behind it. When we position ourselves behind a cat, it will thump its tail repeatedly against us. At times this indicates irritation.

■ Tail wagging – this ranges from small irritated flicks which may express emotional conflict, or a cat torn between intentions, to its most pronounced form between male cats preparing for a fight, accompanied by growling.

■ Tail held down, with elevated rump – when an aggressive cat stands side on to another cat.

■ Fluffed up, arched tail – when a cat stands with an arched back and is torn between being aggressive and defensive. Young cats in particular behave like this towards dogs. The same fluffed tail, but positioned straight out or down, is seen when the balance has tilted towards aggression.

■ Tail down wrapped against body – occurs when a contented cat is lying in comfortable conditions. However, in an intimidated cat this is normally a defensive (but not submissive) posture.

■ Tail vertically in air, cat on its back – the ultimate defensive posture of the cat, protecting the nape of the neck and ready to kick the opponent. It normally follows from the previous position.

Scrapping

Left: The aggressor moves forward, flagging his intentions with ears rotated into the aggressive position, and stands sideways on to the other cat to appear more threatening. The defending cat waves a warning paw

Below: The defending cat on the ground has her ears flattened into the defensive position and repeatedly swipes the air with her forepaws, with claws extended. She puts up a defensive back paw to ward off the attacker

For those who have not kept cats before, some aspects of a cat's aggression can come as a surprise. Once hierarchies are established among dogs, it is not common for dogs to attack bitches. For cats, though, territory is more important than hierarchy, so male-female battles will occur. However, the torn ears of the wide-ranging, intact tom testify to its more frequent and serious fights with other males.

Owners may become concerned when cats that generally get on well suddenly have a spat. One cat can be apparently minding its own business snoozing, when the other cat attacks it seemingly without provocation. In reality, this will not usually come as too much of a surprise to the attacked cat, for despite its appearance it will usually have had its eyes open a little and been aware of the other cat's approach. Indeed, its wariness may be partly responsible for inducing the attack.

It is usual for the defending cat to intercept an attack by rolling on its back and raising its front paws, while its ears flatten instantaneously to protect them. While staying on its back, the defending cat will try to take the weight of the attacker with its back paws and kick to shift it. At the same time, it will initially hold the attacking

cat with its front paws; then, using alternate paws with claws protruding, it may bat rapidly at the head of the aggressor until it withdraws from immediate contact, the defending cat's claws flashing back and forth in front of its nose.

In a battle between an aggressor and a defender, the aggressor is likely to move around the recumbent cat in an angled position. In part this is to appear larger, but in a pause in the scrap it will also be judging carefully the moment when it can again leap onto the other cat. In such a scrap between some household cats, the gaps between clinches are longer than in full fights – lasting perhaps 15 seconds or so, rather than being almost subliminal. During the gaps the aggressor will swish its tail, while the defender on the ground may try a few more front paw bats if it thinks the other cat is coming too close. The aggressor may seem inhibited at times, sitting or giving itself an anxiety lick. However, it is likely to push home its advantage and walk forward towards the other cat's head, with its own head at an angle, aligning itself ready to leap on.

In the face of this, the defending cat, with ears flat, will normally try more front paw bats and may spin its body to disadvantage the attacker, who will try for a head-to-head alignment and a chest grip with its front paws. By turning, the defender can temporarily remove the threat of the aggressor raking with its back legs. The defender will hold its back legs ready to kick off the aggressor. As it leaps on, the aggressor will need to bring its head in close to the defending cat, simultaneously aiming for a chest hug. As it does so, it flicks its ears back from aggressor threat into flat protection positioning against its head.

If the defender can now spin on its back, it can position the other cat so that its back feet land on the ground, not on the defender's stomach, and its head and shoulders will also hit the ground. This

INHARMONIOUS HOUSEHOLDS

With cats that live together, especially if they were introduced too hurriedly, life can be a continuous battlefield. All too often I encounter people whose lives are disrupted by intransigent cats, where one cat is allowed to roam upstairs and another downstairs.

All-out fights between cats disputing territory is not the only form of aggression. Most aggression seen by owners occurs in multi-cat households between cats living relatively harmoniously. This takes place most commonly in the anticipatory build-up a few minutes before feeding time. It usually just takes the form of a raised paw and a hiss, and will normally occur between the same cats, with the aggressor being the same one each time.

whole move of wrong-footing the aggressor can be achieved in less than a second, and the aggressor will need to get up to re-attack. This gives the defender time to reposition, again spinning around on its back to have its head nearest the aggressor, with its back paws ready to kick it off: the defending cat is now completely the wrong way round for the aggressor.

This upside down position, facing the other cat, looks vulnerable, but is the cat's strongest defensive position. A couple of deft defensive parries like this are usually enough to halt the proceedings. The aggressor will lose interest, the defender will settle down, and calm will return to the household, even if the status quo is a little uneven at first.

Above: The aggressor looms over the defender: both have their ears positioned appropriately. The aggressor thrashes his tail from side to side while anticipating leaping onto the defender

Right: Anticipating the aggressor's leap, the defender performs a 'break-dance' spin, whirling around into a position where she can push off the other cat's attack with raking kicks of her back feet. This is the critical defensive move that 'wrong-foots' the aggressor, deciding the outcome of the fight

The Full Fight

Where two large, intact toms behaving territorially are equally matched, and both are flagging aggression with their ears and posture, beware – the fighting that may follow will be extreme. You will now hear the unmistakable, menacing sound of tom cats squaring up for a fight. They will stand with full male heads dangerously close, almost touching. One or the other will swallow and lick its lips.

Having stood their ground for several minutes, one tom suddenly leaps to bite the other's neck. The attacked cat rolls back to adopt the back-on-the-floor position, and both will roll over, trying to obtain a good grasp around the other's chest. This, plus the intensity of battle, is what distinguishes the fight between two aggressors from that between an aggressor and a defender.

The cats may crash to the ground and roll over and over, each trying to hold the other tight with clutching front limbs, making the belly raking tighter. The intensity of the kicks can be such that one cat is thrown clear over the other's head as in a judo move – but the thrown cat will instantly leap back and be grappling again within a second. This speed of return is in marked contrast to the more inhibited scrap of normally harmonious cats from the same household.

In a serious fight, the cats may suddenly break, but will stand face to face and re-threaten, then leap again. When one stays down defensively it has moved into the role of defender, and at some point the remaining aggressor will recognise this and cease attacking. As cats do not have a true submission pose in a form that leaves it symbolically subservient, and not still defensive, this move from mutual aggressor to defender is the nearest they come to acknowledging that the aggressor has finally asserted his dominance.

The noisy drama of a full rolling fight is an earnest matter for both participants, in which the fur really flies as the cats grapple. Its intensity distinguishes it from most cat aggressive interactions which are normally lower-key

9 Cat Chat

The Sounds Cats Make

In human spoken communication we have many languages, with many varied sound constructions from different parts of the world. Yet, although we now have pet cats from around the world, only the Siamese has a strong 'accent'. Some birds, like starlings and mynahs, are great imitators and hijack sounds into their vocabulary. Cat vocabulary, however, is standard enough, as the onomatopoeic names for cat of *mau* (in Egypt), *meo* (Thailand) and *mao* (China) suggest. Traditionally, the English cat is said to 'mew' but the sound is described as 'meow'.

As a small cat, whose main ancestor is the African wildcat, the domestic cat's vocal range excludes the roaring of the big cats, but of the range of sounds it can make, the attention-seeking 'meow' is most commonly heard. This is often at its loudest when the cat comes back into the house and announces its arrival. Studies suggest there is a real feline vocabulary that can be linked to meanings. Phonetic patterns have been recognised, including mating and fighting patterns, purring, spitting, leaving demand, begging demand and complaint.

Closed-mouth sounds include the most

The most recognised cat sound, the meow, has a number of shades of meaning depending on how it is voiced, but all are made in the 700–800Hz (cycles per second) range. The 'm' of the sound is due to it starting with a closed mouth

THE SILENT MEOW

Paul Gallico wrote a much-loved book called *The Silent Meow*, the title of which is based on sound observation, for there are a number of cats that go through much of their lives virtually mute. They will look up plaintively at their owners and silently mouth a 'meow'.

My old female cat, 'Mr Jeremy Fisher', who reached twenty-five years of age and featured strongly in my first book, *The Wild Life of the Domestic Cat*, was one such cat. For many years she would otherwise interact completely normally, except that when she opened her mouth to meow no sound came out. She was used to travelling in the car on occasional long journeys. However, one day I rashly drove through the centre of London, and on looking out through the window at the traffic chaos she suddenly found her voice – and never lost it again! From that day on she used it as if she always had.

In such cats it is not that they are mute physically, but that they give only the behavioural appearance of meowing, without the vocalisation. This can also be seen in cats that are very vocal normally, but will on occasion just mouth the 'meow'. This is an inhibited form of what has been called the begging demand, which itself is an inhibited form of the 'demand' sound. Owners can cause their cat to repeat this sound time after time by looking at it and mimicking the noise. However, there needs to be a motivation for the cat, and this is usually provided by distance. For example, if a cat sitting outside by the door looks up and sees you at an upstairs window, it will make the inhibited demand. If you then repeat either the sounded or the silent meow, depending on what the cat is making, it will continue.

evocatively feline sound, the purr. They also include the rising trill of greeting, where your cat may meet you slightly arching its back, lifting its tail and sometimes lifting briefly on its front paws. The gentle trill of the mother cat responding to her young kittens with a 'come close' sound is similarly structured. Most of the male's chirping sounds made when a queen is presenting in lordosis are made with a closed mouth. The latter two sounds show similar characteristics; they are both requests to 'come close and make contact'. This reinforces the probability that the rising trill made by your cat as it greets you shares a similar meaning.

A low-key murmur – a quieter and shorter form of the trill – is described as an acknowledgement sound. It is met when you gently put a hand on a near-dozing cat and it gives a small sound response, in which you can almost hear a question mark.

The Meow

The sounds coming from a mouth that opens and then gradually closes comprise the range of 'meows', with a fixed vowel-pattern sequence. However, the emphasis, pace and delivery changes the quality of the sounds. These are thought to be 'meows' of demand, begging demand, bewilderment and complaint.

The various 'meows' develop from the kitten's restricted and intense vowel-pattern 'mew'. Whenever even the youngest kitten is out of the nest it will start up its repeat call of distress, to which its mother will respond by coming to its aid. Pick up and hold a week-old kitten for a second longer than it wants to be held and it will let loose a stream of insistent 'mews'! The full meow is not voiced properly in a kitten under eleven weeks of age, and it seems that some cats never do develop the sound.

Turning up the Volume

It is not just in 'finding their voice' that house cats change, for by interplay with their owners they can modify remarkably the sounds they make. It is not only that we encourage our cats, but that house cats actually note our use of vocalisation and 'turn up the volume'! However, some sounds are changed less by association with us, such as the sound made by two cats head to head about to fight (see page 85). As already noted they make an 'anger wail', where the mouth is open and closed enough to overlay a growl with a vowel pattern. The sexual calls of the queen on heat and her mating cry seem to have a combination of a strained intensity and vowel sound.

In the violent sounds of aggression and defence the throat is held tensely, producing the harsher sounds of growling, snarling, hissing and spitting. The shrieking response to sudden pain also falls into this category, which is not the same as the deep distress moan.

Desmond Morris has suggested that the whole vocabulary of the vocal sounds of cats can be summarised in just six messages: I am angry; I am frightened; I am in pain; I want attention; come with me; and I am inoffensive. However, there are many shades of meaning depending on how and where these sounds are delivered.

The Enigmatic Purr

When we sit in a warm room with a cat purring on our lap we are most contented, and so assume that purring is also a signal of our cat's contentment. The image of a mother cat with a line of suckling kittens all purring away while she purrs back seems the ultimate scene of contentment. But cats do not just purr for people and kittens, they purr for themselves: when they are cosy and warm, when they greet a cat companion, or when rubbing and rolling. Purring can occur between juveniles eliciting responses from adults, and adults reassuring juveniles. It is particularly noticeable when made by queens during pre-mating and mating rituals. Yet if it is just to convey contentment, why don't we, or other animals, do it? Why do cats need this special sound?

In fact, it is not just our cats that purr (although they are more competent at it), but the whole cat family and their nearest relatives, which includes civets, genets, mongooses and, surprisingly, hyenas

– although they look like dogs, they are closer to cats! Across this whole group, some sort of purring can be heard when mothers suckle young. The one feature that seems common to the group is that they are mainly solitary nocturnal hunters.

Most of the small cats that are primarily solitary nocturnal hunters living in dense habitat include purring in their repertoire. However, the purr is also shared by the big cats. Tigers, in the wild and in captivity, produce single-stroke, purr-like sounds from deep in their throats on exhalation. The cheetah produces the true purr, like the small cats; anatomically it is different. The big cats have a more flexible hyoid cartilage and cartilage pads in an area of the throat which is more flexible than in the small cats. This allows them to have a resonant roar but may prevent the double-stroke purr.

Although the cat uses the purr in a range of close-contact encouragement and reassurance situations, it is probable that it originates as a co-operative sound between kittens and mother during suckling. Yet virtually all mammals suckle and very few purr.

As kittens suckle for prolonged periods, the purr can be made while on the teat with a closed mouth. Only the range of murmur sounds, including the purr, can be made in this way, and this suggests that the false vocal chords are involved in making the sound. Purring by their mother and litter mates calms kittens and keeps them together in the nest. This calming effect is also likely to reduce clawing at the mother's nipples during squabbles.

In addition, purring provides kittens with a contact call which keeps them together while the mother is away hunting, and its great value is that it is so quiet that it is not overheard by other predators. The mother cat's stealthy approach to her nest and the quiet reassurance the purr gives her young is in stark contrast to the noisy feeding of cubs within the security of a wolf pack. Even when the mother cat is nursing the

Left: The average domestic cat purrs with a fundamental frequency of 26.3 Hertz (cycles per second) (ranging 23–31Hz), and this does not change throughout the adult life. The trace shown is a length of purring recorded in one second at the larynx, where the purr originates. Purring is louder when the cat holds its mouth open, but typically it is only 84dB just 3cm in front of the mouth. It is lower and quieter than most sounds made by the cat

THE LAP CAT

When cats sit on our laps this is a positive action, as free-living adult cats avoid body contact even with other cats of the same group for most of the time, unless permission is given. In addition to reminding the cat of suckling, and being warm and comfortable, we normally give permission for a cat to sit on us by putting our hands forward. We then reinforce it with stroking, which the cat sees as a combination of adult rubbing and maternal licking. Nevertheless, cats do not like being fiddled with too much, particularly by strangers, and when selecting whose lap to sit on will often choose someone in the group who is not being too demonstrative – usually the unfortunate cat hater!

kittens, which she does for 20–70 per cent of her time without other support, she needs a means of keeping them quiet and safe from attacks, and the purr works impressively.

As well as the reassurance and contact conveyed by purring, there is also a statement of existence: 'I am here, I am here'. This is what you hear when you sit on the duvet and up comes a purr from a nearby unseen cat, or when a vet examines a cat.

Communication by Movement

The sounds cats make are not made in isolation but in the context of a wide range of movements including sitting on us, paddling and purring, blinking and staring and other bodily postures.

When cats meet and greet they communicate by posture and contact, a feline body language. When your cat greets you its tail pops up, and when you put your hand down to stroke its back you are reciprocating and reinforcing this move. We both say, 'Hello'. Cats, being strongly territorial and acutely aware of personal space, have ways of signalling that it is mutually all right to approach, and the tail-up movement is one such (see page 81). This signal is often made with the vocal equivalent, the rising trill. If you put your hand down to the cat's head it will be treated as your 'head', as if you were a cat to be mutually rubbed. If you start the movement but don't fully reach the cat, the requirement to rub will make the cat arch up or stand up to meet your hand. However, it can be a mixed response, for if we do not put a hand down to the cat it will rub around our legs or nearby objects, in a similar way to an unsanctioned cat approach.

Paddling and Purring

When your cat leaps up onto your lap, before settling down it is quite likely to paddle on your chest or lap with a slow, rhythmic open and closing of its front paws, and this will often be accompanied by purring. These are the same

THE WITCH'S CAT POSTURE

Detailed observations of cats suggest that their body postures and facial expressions do not just go from the aggressive threat/attack to the low crouch of standing defence, but that there is also a mid-point of arousal, in the shape of the familiar witch's cat posture. In this the cat stands high on its legs, with its back fixed in a high arch and its tail and back bristling with fur. This is the pose that inexperienced juvenile cats adopt when faced with an apparent threat. The cat also stands side-on to the threat to maximise the visual image it presents. It is often used against dogs.

This posture is a mixture of defence and attack – it has the high legs of the aggressive cat but the pulled-down head, bristling fur and hissing of the defensive cat. As a defensive cat trying to be aggressive, it bravely stands its ground. Its walking movements are wooden, due to the conflict within the cat. Its eyes too are midway between fearful pupil dilation and aggressive closure.

Left: It's nice to feel kneaded! This, like some other behaviour between cats and ourselves is neotenous, that is reliving kittenhood

He may not look hard at work, but he's on duty, keeping another cat out of his domain with a fixed stare

moves that kittens make around their mother's teats between three weeks of age and weaning. Our warm laps and large size seem to bring out this kittenish behaviour in our cats.

However, it is not so cut and dried, for adult queens coming into heat carry out the same opening and closing of their paws while both paddling and purring. To test if a queen is approaching oestrus, breeders may stroke or pat her along her back towards the tail – if she is, she will elevate her rump and adopt the lordosis position. If you do the same to the cat paddling on your lap you will elicit a similar response. During pro-oestrus the queen will be very friendly and will rub around with her head or neck. Consequently, cats' behaviour towards us can be viewed as part juvenile and part sexual, but clearly affectionate.

Blinking and Staring

These two behaviours are the reverse of each other, having opposite meanings. Blinking is a very powerful communication as a reassurance signal and is commonly used between cats when they are sitting or lying in a hunched-up, sphinx-like position. I have used the blink to relax house cats, feral cats and even tigers in the wild. Cats are masters of stealth and to try to creep directly towards them is likely to unnerve them. Instead, it is important to put them at their ease by allowing them to read the signals that tell them you are not a threat. It is vital not to give the impression of approaching and to perform blinks that are slow and definite, with eyes hooded.

One thing that filming cats has taught me is how threatening and unsettling they find a continuous stare. It is certainly something that cats use themselves to that effect. When an aggressive cat is intimidating another with slit-pupil eyes and threatening ears, it gives an unblinking stare. Cats do look directly at each other at times, such as in the build-up to

renewed mating, but then the nature of the eyes is not threatening.

The stare threat is used effectively by cats in maintaining territorial distance. This is commonly seen in the cats in our gardens, particularly in high-density urban areas. Cats will sit for hours on end, sphinx-like with front paws tucked under, and just stare from a vantage point within their territory at another cat within the territorial limits. This is most commonly seen between queens of adjacent territories, and these 'stare-pairs' repeat the procedure day after day. In our interactions with cats, it is important to remember how intimidating and threatening a stare can be. It is vital to diffuse the attention.

10 The Human–Cat Relationship

The relationship between cats and people has not been a static one, but has shown considerable evolution and change over the centuries. The main ancestor of the domestic cat, the African wildcat of North Africa (*Felis silvestris lybica*), overlapped the Marsh cat (*Felis chaus*) in its range in Egypt along the Nile. Although domestication of the cat may also have occurred elsewhere, the only significant mass of evidence for this in terms of paintings, statues, carvings, mummies and bones comes from ancient Egypt, where cats were kept confined in temples. It may be that a cross between the Marsh cat and African wildcat assisted in domestication. Certainly, the formation of small towns and villages with wastes ideal for scavenging attracted African wildcats in the first instance, and food sources near humans was probably a major factor in domestication.

Development of the Relationship

While the cat was kept as a fertility goddess in ancient Egypt and despised as an incarnation of the devil in sixteenth- and seventeenth-century Britain and Europe, for much of mankind's 3,500-year relationship with it the cat has been employed mainly for its rodent controlling skills. Only with the advent of cat showing, following the first major cat show inaugurated by Harrison Weir at the Crystal Palace in London in 1871, did the cat become acceptable as a pet to a wide spectrum of the population. Its spectacular rise to become probably the most commonly owned pet in the world is therefore quite remarkable.

When we hold our cats and confide our innermost secrets to what is often our closest companion, this intimate act is being repeated all around the world. Yet just a few generations ago such universal devotion to cats would have been unthinkable. Our social changes to accommodate the cat have been dramatic, but no less than the huge leap taken by the solitary hunting cat to become a pet.

Social Structure

In essence, we behave in our relationships to cats as if they are people, and cats behave in many ways to us as if we are cats. Not all pet keeping is like this; owners of exotic collections of tropical fish do not generally have that response – theirs is more the passion for fascinating shapes. But both cats and dogs are mammals of sufficient size and response to be companionable.

Due to their wolf ancestry, dogs have a hierarchical social structure. They are secure with the owner as the dominant partner and what we regard as affection from the dog is largely appeasement behaviour. The social structure of cats living feral within a group is not primarily hierarchical, therefore we have a relationship with our cats which is on a more equal footing. In their looser social structure, cats have no need of appeasement behaviour. However, there is a blurring of roles in our relationship with cats, for sometimes they behave towards us as adult cats, sometimes as if they are juveniles, and sometimes as if we are sexual partners!

INCREASING POPULARITY

As you walk around the streets of Britain you will find a cat at virtually every other house, sitting in the doorway sunning itself. In North America the indoor/outdoor cat situation has changed dramatically over the last twenty years, and most cats are now confined. There has been a massive increase in cat ownership in North America and across Europe in recent years. In Britain in the early 1980s there were around 5 million owned cats but by 1993 they had risen to 7.1 million, pushing the dog out of the number 1 slot. In the USA, this happened earlier, in 1987, when the cat reached 56.2 million.

This 30 per cent increase is largely due to changes in our lifestyles. Professional couples who are both working find the regular tie of taking a dog for a walk too restricting on their timetable. Dogs are less convenient for urban living. Young couples who are delaying having children while furthering their careers are keeping cats instead. However, because they go out to work many people have a second cat so that the two can 'keep each other company'.

Although the number of owned cats rose throughout the 1980s and 1990s, and this was matched by a similar increase in cat-owning households, the number of cat-owning households remained lower than that of dog-owning ones, although these are expected to meet soon. This simple change has dramatically affected the dynamic of many cat-owning households.

How Your Cat Reacts to You

Cats living together in feral groups do interact affectionately. Similarly, in multi-cat households adult cats may greet each other with head rubbing, and they may lie together and groom each other, although such interludes are often tempered with the odd spat.

In contrast, affectionate interactions occur extensively between domestic cats and the people with whom they live. Why should this happen so much more between cats and people than between cats and cats, when we are different species? Paul Leyhausen suggests that we are similar enough to evoke the cat's juvenile behaviour, but not similar enough to provoke defence and/or attack behaviour. Juvenile behaviour goes through developmental stages in kittens, transforming into adult behaviour as the kitten matures. We certainly are the focus for juvenile features of behaviour, from purring to play, but why should this be?

It is partly due to our huge size relative to an adult cat, which is not that different from the proportion of an adult cat to a young kitten. By sitting on our lap the cat receives as much contact and warmth as it would have from its mother – small wonder it is triggered into juvenile behaviour. When they stroke the cat, our hands are proportionally similar in size to its mother's

Cats interact more with people than with other cats. For the cat, stroking is an echo of being groomed by its mother

tongue when it was a kitten, and our stroking movements may remind the cat of the washing it received from its mother while it was young.

Early Conditioning

Why should we evoke positive rather than negative social interactions with cats? The reality is that cats that have not become at ease with people do not prolong or seek contact. Some feral cats may habituate to someone who feeds them over a long period of time, while others remain distant. We normally have much closer relationships with our house cats, and early conditioning of domestic kittens to people (see Chapter 6) allows them to accept people in a more relaxed way. Owners reinforce this, normally holding their cats at some time on most days; usually at some point picking them up and carrying them around, a task previously only done by the cats' mother.

Most feral groups consist of related family members and close social bonding provides advantages for the group. Reactions are generally more amicable towards familiar cats of the group than towards strangers. In my Fitzroy Square group (see Chapter 4), two adults who had been litter mates shared the same basement retreat and often sat together. Such closeness is not uncommon between mother and daughter, but in this case one of the cats was male. If they had been people, any outside observer would have called them friends. To our own household cats we are members of a common group, and with cats that were habituated early towards people at the kitten handling stage we are accepted even more fully.

Domestication has been followed by selection for more juvenile behaviour in cats, not just for them to be more tractable, but also to be more trusting of people. Placing ourselves in the position of the hunter, the provider of food, also reinforces our mothering role and the cat's kitten role.

Friendly Intentions

When cats meet they often investigate each other briefly, nose towards nose. If they are of the same group this may be followed by a head rub, then

which we are sitting. In contrast, nervous cats, despite their commitment to their owners, often remain intimidated by our large, relatively flat faces and reserve the 'head rub' for our hands.

Cat Character

Ask any cat owner and they will say that their cat is an individual, with its own definite character. Any owner who has had many cats over the years will remember them all as having distinct characters, and will recall stories that demonstrate a particular aspect of their old friends. Yet words like character or friend are

Above: Even without thinking about it, we bond positively with our cats, stroking them and providing the companionship of a larger animal. No wonder they behave like kittens towards us

Right: A child's show of affection can be robust, but contented family cats usually show remarkable tolerance

walking in contact with the other cat's side. However, with strange cats the response is much more likely to be cautious and wary, with ear and facial movements flagging intentions. When we greet a cat they focus on and rub against our hands, instead of our head or nose, having first sniffed a finger. Our hands do not have ears to convey messages and we fairly quickly make the positive move of stroking the cat. It is likely that our success at socialising with cats is due partly to our inability to 'flag' aggressive intentions with ear movements. A major factor is that we both initiate and sustain prolonged affectionate moves, and this is sufficient to foster commitment to us in our cats.

Many 'laid back' cats are happy to head rub their owners when both heads are at the same height, such as when a cat is held in our arms or walks along the back of the sofa or armchair in

rarely used by behaviourists, although interestingly they do use human terms to describe more negative features, such as aggression.

But whether we call it style, personality or character, owners know such features can be identified in cats. I often say of Leroy, my present tom, that he is 'laid back', 'Mr Cool', but a 'demon hunter'. In contrast, Tabitha, the female, is 'more nervy'. My previous house cat, who lived until she was almost twenty-five years old, was always 'a lady'. For me, such terms encapsulate the essence of their behaviour as I knew and know them.

How Your Cat Reacts to You

Experience and Character

Whether we are cat or person, we are given a pack of genes at conception, but how we behave and respond during life depends also on our early developmental experiences. When we interact with other people, we come with a baggage of preconceived ideas and assumptions, mostly gathered during early experiences. We rarely look at an event unclouded by the past. It is the same with cats.

How do you really look at your cat? All cat owners recognise that their cats have a distinct personality or character and how your cat reacts towards you, other people and other cats is largely dependent on its character. Is it affectionate and outgoing, or nervous and shy? A cat's character, in broad terms, remains with it for most of its life. Its very constancy enables us to recognise a particular character trait. Aspects within that can change, as factors around the cat change. It can develop what we term 'behavioural problems', usually due to an increase in local cat density. Nonetheless, our cat is still recognisably 'our cat'.

We treasure our cat's friendliness and shows of affection, yet these are in large part a result of early kitten socialisation, later habituation and litter birthweight. How much of a cat's character is not learned but innate is not yet known. This nature or nurture argument has long been debated in studies of human personality as well.

Kittens demonstrate distinct patterns of behaviour which are recognisable as character traits as soon as they begin to move about in the nest. Some are more demanding, outgoing and confident, while others are more retiring, nervous and shy. Yet even if reared by a mother who is well integrated and affectionate towards people, kittens at very early stages that have not been socialised to accept people are not initially affectionate towards

them. Kittens that have been habituated to humans are confident enough to approach new objects more quickly than non-habituated ones, so although young kittens may have a particular

Cats' characters are formed early, but spending more time with them improves socialisation, and is rewarding for all ages of the family

nature genetically, early socialisation dramatically affects the degree of warmth in their relationships with humans.

Some people complain that cats are 'stand-offish', but the more time owners spend in the company of their cats (allowing the cat to make the contact), the more interactions occur and the 'friendlier' the cat will appear.

How We Talk to Cats

It seems that we are nearly all Doctor Dolittles in that we talk to the animals – at least the vast majority of us talk to our cats. A survey by Peter Borchelt and Victoria Voith of nearly 900 cat owners visiting four American veterinary hospitals found that 96 per cent of household cats were talked to at least once a day. Most of the owners felt that they and their cats communicated their moods to each other. Three-quarters of owners felt they were either usually or always aware of their cat's moods, while over half believed their cats were similarly aware of their moods. Around 65 per cent recognised that they talked to their cats entirely as if they were some sort of person, mostly a child. An unassailable 99 per cent considered their cats as members of their family. When talking to their cats only 13 per cent did so as if they were just pets. Despite that, half of the owners thought of their cats as animal members of the family. Nearly half of the owners surveyed would at least occasionally confide their problems or matters of importance to their cats by talking to them. So what do we say to our cats, and how much is it a conversation and how much a monologue?

Conditioned Communication

It is easy to overlook even in communication that we are part of the equation, and what we do affects our cats' responses. In any household there is a mutually understandable cat and human pidgin language that has arisen from a joint need to communicate. Yet the conditioning is two-way. The surveys suggest that most of us use a manner of speaking as if to children, and at times such as feeding virtually everyone speaks to their cat as if talking to a baby, slightly higher pitched, with sing-song rhythms. We tap the plate and use exactly the same wording and intonation each time with phrases like, 'Do you want something to eat?' and 'Here Kitty, Kitty, Kitty'. Our movements as well as our sounds are repetitive and recognisable as we pick up the saucer and tin opener.

As urban, free-living feral cats come to recognise and anticipate our regular timetable, actions and sounds linked to feeding, it is no surprise that so do our house cats. Other regular patterns such as sitting down after meals, when the cat may leap onto your lap and settle down for a nap, are repeat sequences that cats recognise and interpret. Similarly, cats not only recognise that we

Cats are companionable with their householders. Where you regularly sit quietly to read or write, so will your cat

have a behaviour timetable from waking up to going to bed again, but regular activities like returning from work are anticipated. Where cats are rewarded with attention, being there for the owner's arrival and recognising the car show an accurate gauging of time plus an interpretation of engine noises. In this they are masters of our body language and sounds.

Do You Sleep with Your Cat?

Your cat's favourite snoozing spots, both inside the house and out, carry its scent; its unchallenged state gives it the confidence to rest, and our scent conveys similar support

Some people wouldn't dream of going to bed without a mound of cats all over it, while other owners wouldn't contemplate such a pet invasion. There can be a very practical reason for the pet embargo here – fleas in the duvet! However, even when the fleas are controlled, choices are not random. A recent survey by Barrie Sinrod in the United States of 10,000 pet-owning households found that 60 per cent of pets sleep in or on their owner's beds. The group most likely to cuddle up with their pet is young women between 18 and 34 years old. In contrast, married men over 45 tolerate the pet only if it sleeps on the foot of the bed.

Our reaction to our pets on the bed is not as if they are inanimate hot-water bottles, for owners commonly comfort themselves around their pets so as not to disturb them. Given the small size of cats relative to people it is remarkable that they aren't squashed, but somehow people move even in their

sleep to allow for their pet. Due to their love of getting into things, given half a chance most cats will try to burrow under the duvet and snuggle up close to their owner. Many owners draw the line at this, particularly when they want to sleep, and will haul the cat onto the top. The cat usually prefers to take up a position in the hollow at the back of the legs if stopped from lying directly on its owner.

Cats appreciate the prolonged, restful and close contact they can achieve by sleeping with us. However, beds are used for more than sleeping. The Sinrod study found that 73 per cent of American pet owners had sex while their pet was in the room. This demonstrates the privileged role pets play in our lives, for while we don't view our pets as inanimate, and in many ways relate to them as members of our human group, we don't normally allow other family members to be present at such times! It can't be that we think cats are oblivious to what is going on, for they often demonstrate empathetic behaviour, aroused by the heightened social interactions and reinforced by our sexual scents. We should not be surprised by our sexual scents having this fascination, when a cat's dramatic response to the plant nepeta (catnip) is considered (see page 43). Much of their interpretation of their own feline world is via the sexual scent of other cats.

Although our noses are not in the same league as those of cats and dogs, our behaviour towards our cats is often quite blatantly sexual in origin. In Sinrod's survey 81 per cent of cat owners were found to kiss their cats – nearly 20 per cent more than would kiss their dogs. Due to the softer nature of their fur we stroke cats for prolonged periods, while we tend to pat dogs. Stroking, though, is something we predominantly reserve for

sexual partners, tending to restrict its use within the family. However, our stroking of a cat is reinforced because it is reciprocated. When we repeatedly stroke a cat on its back, it raises its rump and tail and lowers its forelimbs. This is a mixed message: in part the raised tail greeting, but also a sexual statement.

Our own sex will usually have an effect on our relationship with a cat. The long-standing feline/feminine relationship which dates from ancient Egypt (see page 8) also operates today. Many women believe they empathise more naturally with cats. Whatever the basis for that belief, women are still more likely to feed cats, go down to their level, and speak gently and more frequently to them. Single women spend more time interacting with cats than those living with partners.

THE IMPORTANCE OF SCENT

Take your clothes off and hop into the bath, and before you can sink into the water a cat will be settling onto your still warm clothes! Our cats like companionship and will follow us about in our routine, and discarded clothes do make a comfortable bed, but there is more to it than that. The warmth is attractive, and soft or woollen clothing can elicit foot-paddling, but a key feature in the attraction is our scent. As understanding the scent of fellow cats of a group is important to feral-living cats, so too is an understanding and appreciation of the scent of their

fellow 'cats' to house cats. Its unchallenged strength is reassuring to our cats.

If you allow your cats into your bedrooms you will find a similar preference for the bed you sleep in (even when you are not there) over other beds. Your use of the bed and your scent are the attraction. Other factors can make other beds attractive, such as a patch of warm sunlight or an adjacent radiator, as well as the cat's relationship with other members of the family.

Cats gain territorial confidence and reassurance by our presence and scent in the inner sanctum of our bedroom, where they can rest secure. Consequently confined cats are less likely to have stress-linked behaviour problems if allowed to sleep on their owner's beds

11 Cat Naps

Cats have sleep down to such a fine art that the ability to snatch a moment of sleep is actually named after them – a 'cat nap'. Cats can readily sleep for up to eighteen hours a day, yet when their sleep time has been studied in a confined, captive state they have been found to sleep for only thirteen hours.

Sleep and Lifestyle

The proportion of sleep per twenty-four hours relates to the lifestyle of a species; for example, the incredibly slow-moving sloth can spend over 80 per cent of its time asleep, while the small hunting shrew has to eat every two hours or die and so hardly sleeps at all. However, although the cat is also a hunter, its larger size means it is not hostage to sleep loss!

While large grazers, from horses through to elephants, sleep remarkably little – perhaps four or five hours in twenty-four – as they need to consume huge quantities of grass and leaves, the protein-rich diet of the cat allows it to invest more heavily in sleep. This probably enhances the cat's longevity, which is greater than would be anticipated from its size alone.

Not only does the cat, with its higher protein intake, sleep more than the dog but as a lone hunter it does not have the group support of the dog family. Consequently, it needs a device that will allow it to doze, but then be awake in an instant. The cat has a third eyelid, the translucent nictating membrane, and when the merest shadow crosses it the cat springs into alert action. Usually the cat has its outer eyelids partly closed so that the eyes do not look strange, but when this is not so, owners unused to seeing completely white, pupil-less eyes may become alarmed.

Resting, but carefully keeping an eye on what's about (but keeping its nose covered, see page 102)

The Significance of Sleep

A kitten spends most of its first week of life asleep, but even when adult a cat spends more time asleep than most species

Sleep as Defence

As the mother of a new litter has to leave the 'nest' in order to catch food, it is vital in the early days that the young do not move from this safe haven. Consequently, kittens are born in a more immature state than many mammals, with the result that while young they sleep for a proportionally greater part of each twenty-four hours. This is primarily deep sleep, which lasts for about twelve hours in every twenty-four. After their first month of life, kittens change over to the adult pattern (see page 65).

DOZING PLACES

Cats like to feel confident before sleeping and nervous, dependent cats will often accompany their owners around the home for part of the day, and sleep near them.

Some cats are most particular in their requirements for cover in the garden before they will attempt to snooze. Just as airing cupboards can be favourite warm sleeping spots inside a house, so a cat will have definite warming and cooling dozing places in the garden. The resident cat will become very upset if another cat usurps its sleeping spots (see Chapter 8 page 80).

The Significance of Sleep

Sleeping Position

The sleeping postures that animals adopt depend very much on their size and body shape, as well as on environmental temperature. Cats share their sleeping postures with other carnivores: they either lie down, crouched sphinx-like with their heads down, or lie on their side, with their bodies curled around to varying degrees depending on the temperature.

When a cat is hot it will lie stretched out long in order to increase heat loss, but in cooler conditions it will curl itself up, sometimes to the extent of covering its nose with either a paw or its tail. The image that advertisers like to project of the contented cat in a room warmed by a particular fuel is therefore usually wrong, in that it shows a curled cat!

Sleep Patterns

For most animals, sleep patterns follow a biological clock, a circadian rhythm tied to the natural twenty-four-hour cycle of the earth's rotation. Predators generally restrict their waking activities to the times when they are most likely to encounter their prey. In most parts of the world,

awake | light sleep | asleep

Typical sleep trace pattern

and especially in the northern temperate lands, small mammals like mice and voles, that are warm-blooded and therefore not directly dependent on the sun, restrict their food foraging until after dark, as they are vulnerable to numerous predators during the day. As a consequence, cats developed night vision and nocturnal hunting skills. This has given them a natural sleep rhythm whereby they spend a major part of the daylight hours snoozing, to allow for the alertness needed for their nocturnal forays.

In some circumstances this behaviour may be reversed. For example, Australia is the only continent where reptiles make up a significant proportion of a feral cat's diet. As reptiles need the sun's warmth in order to function and are therefore active during the day, the hunting and sleeping timetable of the cats that prey on them is adjusted accordingly.

Unfortunately, a cat's sleep pattern does not necessarily coincide with our own, as you will have noticed if you have been rudely awakened at five o'clock in the morning by an insistent meowing after your cat has returned from a nocturnal foray. In addition, the cat does not take all its sleep in one long stretch, as we generally do, but will snatch a number of cat naps as well.

Above: Dozing in the early morning sun, but with paws tucked in to preserve heat
Left: A contented cat lying out long in front of a hot kitchen range
Below left: Settling down on a favoured warm snoozing spot in the home

WHAT HAPPENS DURING SLEEP

Although a cat may snatch a cat nap, when settling down to sleep it goes through a period of light sleep, followed after half an hour or so by deep sleep for a shorter period of about seven minutes. During light sleep the cat can be woken easily. As it moves into deep sleep, the pattern of its brain waves more closely resembles a wakeful state. The cat drifts off into periods of deep sleep alternated with periods of light sleep. In total, it spends about 30 per cent of its sleeping time in deep sleep.

During periods of deep sleep the cat's eyes show rapid eye movements. This is called REM sleep, and the same thing is observed in people. In us, it is during the deep sleep REM period that we dream, and it seems logical that cats too dream during this period. Just like us, they twitch – their paws, ears and mouth area may twitch rhythmically despite the rest of their body being completely relaxed. People often mumble during this period of sleep and cats similarly tend to mutter and grunt.

Waking Up

Yawning

Like us, cats usually yawn widely on waking, but between cats and people, yawning has further significance both as a reassurance signal and, inadvertently, as a greeting. If you walk into a room and approach your cat, who has been cat napping, your sudden appearance can cause it to give a yawn of recognition.

Right: The tongue curls back and the whiskers space out as the muzzle tightens in a full yawn

Stretching Exercises

When your cat wakes up it will give a huge yawn, with its mouth fully open and its tongue curled into a ladle shape. It will often stretch its paws out long as well. If it doesn't just settle down for another well-earned snooze, but is actually rising for something important, such as a snack, it will stand tall on long straight legs, pulling itself together while arching its back high, raising and rippling its muscles clearly. It will then move forward, but only to go through the next stage of the waking exercise ritual with the long forward stretch. In this, the cat's rear end stays up and is pushed back while the spine curves down in an arc, with the head held low and the front legs and paws extended forwards. The cat then walks and leans forward, with the thoracic spine pulling against its back legs, which it stretches out long as it did with the front paws earlier.

This set of isometric waking warm-up exercises allows the cat to remain in tip-top condition. Through a combination of standing, arching and stretching, it restores flexibility to its spine. This flexibility is of fundamental importance, as not only does it allow the cat to increase its stride length when running, it also allows it to wash and groom itself all over.

Below left and right: The cat's way to stay supple and avoid a bad back. Its stretching routine, performed on waking, systematically pulls on the vertebral column, arcing up and down

Washing and Grooming

On waking, a cat will often give itself a wash and a groom. When we stroke or groom a cat we remind it of kittenhood, when all grooming in the first few days of life was carried out by its mother. As she licked her kittens, she not only groomed their coats but stimulated them to urinate and defecate.

Grooming remains important throughout a cat's life. Although we say that cats just eat and sleep, they have been found to spend from one-third to one-half of their waking hours in grooming. No wonder they become exhausted and have to spend more time sleeping! Cats shed some hairs right through the year, but this is heaviest in the spring and most noticeable in fuller-coated cats. Even so, all cats have to invest much time in coat care, especially compared to some other animals, and certain factors make such fastidiousness possible.

The cat's incredibly flexible spine allows it to wash the middle of its back

Cats groom much of their coat directly, but wash their faces by dampening the inside edge of their lower foreleg and wiping with that

Aims and Equipment

Successful grooming depends on the cat having the right equipment and using it well. Cats possess incredibly supple bodies, which allow them to reach nearly all parts easily. They can wash around the urinogenital area effortlessly by sitting down with either a balancing back leg raised, or back legs splayed. With a more concerted effort, they can even reach the middle of their backs.

The cat's tongue is its most important grooming tool. The spines on its surface (see page 21) act as a very effective comb, and after a thorough grooming session you will find a mat of hairs on

MUTUAL AND SELF-GROOMING

Behaviourists classify self-grooming as autogrooming to distinguish it from mutual grooming (allogrooming). At birth, the mother's tongue removes the birth membranes and begins to groom the kitten, and it is this vigorous washing that stirs it into life after she has washed away the mucus blocking its nose and mouth. All through kittenhood it will be washed by its mother, but at around three weeks of age it will begin to groom itself, and by six weeks this has become a more frequent activity. By this time the kitten may also be mutually grooming a companion in the litter. Cats who live together from kittenhood tend to retain this grooming bond and they will groom each other from time to time. This will also happen with non-related household cats.

The closeness and responsiveness that grooming can produce is seen in free-living cats before mating, when the male seeking to make advances will groom the back of the female's head when permitted. With neutered house cats, allogrooming can be appreciated and social bonding increased, but the male may find the female nervous and be on the receiving end of a well-timed blow.

your favourite armchair. This is the amount the cat has managed not to swallow. Swallowing hair in itself is not a disaster, as cats will vomit up hairballs periodically; however, some cats are prone to internal impaction of hair and for these additional help in grooming is useful.

Grooming can be a problem for cats with long hair, as they have to pull their tongues through each section of coat for a longer period of time. Unfortunately, modern breeding of Persians has produced a coat that does not self-align as well as the traditional longhair coats, and for these cats daily additional grooming by their owners is vital.

As well as acting as a comb, the cat's tongue makes a good sponge. The papillae on its surface increase the volume of liquid held there and this is useful when a lot of saliva is required for grooming. This feature is also important in temperature control. The cat has watery sweat glands only in localised areas such as the pads of the feet, so it can avoid wet fur which would cause over-chilling. In hot weather, the cat compensates for this by using its tongue as a combined sponge and brush, and the even spread of saliva can increase evaporation cooling by up to one-third. In

cold conditions, the brushing effect can also be effective in maintaining body heat, because a fluffed-up coat traps a layer of air which acts as an insulator. A tangled or ruffled coat does this less efficiently.

Grooming also enables a cat to spread its own scent around its coat, while at the same collecting taste information on things with which it has come into contact. When we handle a cat, it will often fussily wash its coat back into place, thereby re-establishing its own scent. The violent washing following mating is in part for a similar reason (see page 57).

Grooming is also an essential activity to control and reduce infestations of fleas. With a burden of biting fleas, the cat will suddenly swing around violently and begin intensive sucking grooming on its back, or just as immediately sit down to attack the area near its tail. This will not necessarily lead to coat loss.

A simple quick lick to a leg will not develop into grooming; it is a displacement activity that cats frequently use in response to the smallest increase in anxiety

12 How Cats Think
Instinct and Intelligence

While Descartes is celebrated for his assertion that 'I think, therefore I am', he was less flattering towards animals and their behaviour, regarding them as merely automatons. In that, he agreed with the dogma of mainstream European Christianity unchanged since the time of St Thomas Aquinas, who propounded the theory that animals were devoid of free will. To keep themselves 'pure' from anthropomorphic interpretations of animal thought and judgement, many people – including numerous twentieth-century biologists – have been content to attribute the functioning of animals to 'instinct'.

Darwin was more rational, and suggested that instinct could be thought of as acting in a reflex way. He argued that 'animals possess some power of reasoning', and that 'the difference in mind between man and the higher animals, great as it is, certainly is one of degree and not of kind'.

Long-lived mammals have lengthy periods of development after birth. As kittens are born at an early stage of development compared to some

INTELLIGENCE EXPERIMENTS

Unfortunately, much of the debate over instinct or intelligence has foundered on the obsessive belief that it must be all one or the other. In addition, the methodology used in trying to unravel the tangle has been suspect. Like us, cats use a mixture of both instinct and intelligence. In 1911, the American psychologist E.L. Thorndike published his book *Animal Intelligence*, in which he described using puzzle boxes to test intelligence. Cats and other species were placed in these boxes and had to push levers or pull string to escape. He described their approach to getting out as 'trial and error' and the phrase has stuck. His interpretation of his findings was that a cat's approach to the problem was quite 'mechanical', and that random success was then adopted. His interpretations, and in particular his non-cat friendly experiments, have since been questioned: pushing levers in this manner is quite inappropriate to the cat's way of life. Gradually, generations of laboratory tests have given way to observing animals in their natural environment.

other mammals, they have a relatively long rearing period within which instinct can be tempered by environmental influences and a period of learning. Thus learning plays upon the inherited instincts and can fit a cat better for the conditions it will meet. Most significantly, this has enabled cats to integrate better than might have been anticipated for relatively solitary carnivores. The period of litter-mate socialisation has been highjacked by owners of cats – this was even more so in the past, when house cats were brought up in houses rather than catteries.

Yet despite this, dog owners and designers of animal IQ tests maintain that cats score badly compared to dogs. However, this is an unwarranted slur, as such results were simply reflections of the trainability of dogs, measured against the unresponsive nature of cats in the same situation. This in turn arises from the biological disposition within dogs to form a group or pack. Such behaviour is not relevant to the cat's more solitary hunting lifestyle, where it has to make individual evaluations of a situation and then act on them. To be instructed what to do is meaningless for a cat. Yet when they are self-motivated, cats do accomplish similar 'tricks' to those of dogs – opening door latches, following elaborate routes, and so on.

Interpretations of how cats use their territories have changed. The model of 'patrolling the boundary' has been superseded by the idea that cats spend more time in areas in which they are

confident, due to marking. This model suggests that a limited number of inherited imperatives produce the basis of territorial behaviour in cats:
1 Stay in areas in which you feel more confident.
2 Travel more if you need more food.

Although inheriting general rules such as instincts could provide a framework for behaviour, a real landscape is made up of features with changing parameters, so the ability to learn and interpret gives flexibility.

The Feline Brain

Cats' brains grow rapidly through kittenhood, so that by three months when they gain an adult body size of 20–30g (¾–1oz) the brain is five times larger than when they were born. It has often been stated, simplistically, that the larger the brain the more intelligent the animal. However, larger animals also have larger bodies that require more neurological 'wiring'. Cats, as carnivores, do have a proportionally larger brain-to-body ratio than do rats or mice, so they should have the advantage. This is because larger brains not only have more brain cells, but also many more connections between those cells.

Structure of the Brain

The cat brain consists of three main sections: the forebrain, midbrain and hindbrain. If the brain is viewed from above, the visible surface is dominated by two structures: the cerebellum of the hindbrain, and the cerebral cortex of the forebrain. Like that of most mammals, the cerebellum of the cat is highly convoluted. It is proportionally larger than that of most mammals and specifically controls the co-ordination of movement, balance and posture – vital for a tree-climbing predator.

The surface of the cat's forebrain is significantly

The brain of the curious inquisitive cat produces an instinctive reaction, blended with learned skill, as excellent eye and paw co-ordination deliver a deft strike

folded into a series of ridges. In large mammals like whales and man this cortex surface is highly folded, while in smaller animals like rats or rabbits there is very little folding. New-born kittens have little folding of the neocortex compared to adult cats. It is therefore not surprising that lack of adequate nutrition during kittenhood can seriously damage cat behaviour patterns.

The surface of the cerebral hemispheres has been mapped into areas that receive information from sense receptors, and areas that control movements of the body. Generally, the larger these areas are relative to each other, the more brain cells are involved with that function, which reflects its importance to the animal. The nocturnal hunting cat, with its ears finely attuned to the small sounds made by wood mice at night, has a proportionally large area (25 per cent, compared to 10 per cent for a rat) dedicated to receiving and interpreting aural information. As the sense of touch originates from the skin, a map of the cortex area for the relevant sensory neurones reveals a distorted picture of the cat – areas like the head and tongue that require greater sensitivity and have more nerve endings are correspondingly larger.

Instinctive or Learned?

The part of a cat's brain that is connected with pleasure and avoidance of pain is the limbic system, which encompasses parts of the upper end of the brain stem and the base of the cerebrum. It is strongly involved with emotional states.

In the past, psychology researchers working with cats found that by stimulating the hypothalamus they could cause them to retract their ears, crouch, growl, raise their backs and lash their tails in a stereotyped reflex manner. Similar stimulation of the hypothalamus of cats that normally did not attack rats would initiate an attack and result in the killing of the rats. However, the attack lacked the skill and refinement of a learned approach, being instead mechanically direct, and had the appearance of an instinctive behaviour pattern.

The older, less elaborate parts of the brain are the underlying extension of the spinal cord consisting of the hindbrain, midbrain and lower forebrain, and these are the most similar parts of the brain in all mammals.

Electrical stimulation of appropriate areas of the hypothalamus seems to release the instinctive behaviour that is innate in the animal. Basic functions like eating, drinking and copulation are all mediated in this way. The behaviours of dramatic emotions of rage, aggression, fear and so on are under this same control.

All this implies that we can distinguish between instinctive and learned behaviour by having sites that appear to control such behaviour in separate parts of the brain – the instinctive in the older brain and the learned in the newer, developed brain. Equally implicit is that the behaviour of dramatic emotion is in an ancient part of the brain which is remarkably similar in us and cats. Consequently, it is in the grand passions or emotions that our feelings are identifiably most similar to those of the cat.

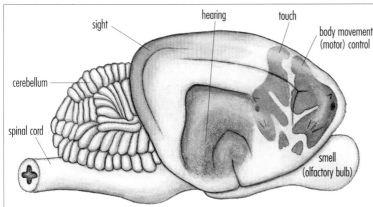

The surface of the brain with positions of sensory and motor areas of the neocortex of cat's cerebral hemisphere

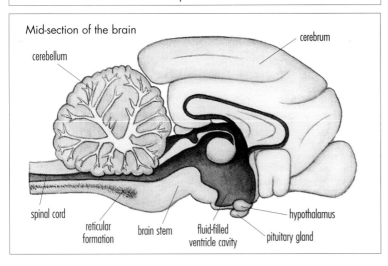

Mid-section of the brain

Hormones and Behaviour

Many of the physiological events in the body during emotional reactions are mediated by hormones such as adrenalin, which is produced by the adrenal glands. A series of glands in the cat's body release their hormones into its bloodstream, from where they have a general effect on the body. Adrenalin is popularly described as causing a 'fight or flight' response. It mobilises the cat's anxiety, fear and aggression. Similarly, testosterone released by the testes affects the behaviour and build of a tom cat.

The link between hormone control and the brain is the hypothalamus, which triggers the release of trophic hormones from the pituitary gland. This causes the other hormone glands of the body to release their hormones. Because of this ability to regulate the other glands, the pituitary has been called 'the controller gland'. The circulating levels of hormones feed back to the hypothalamus and pituitary gland, which adjusts production of the trophic hormones as necessary.

The Nervous System

The cat's central nervous system consists of the controlling brain and its main conduit, the spinal cord. From here a network of nerves runs through the body, carrying impulses to the muscles and sending back sensory messages to the brain.

The more instantaneous reactions of the body – the flight/fight situations – are controlled by the autonomic nervous system. Involuntary muscles such as those of the cat's eyes have a double nerve supply, with matching nerve endings in both the parasympathetic and sympathetic nervous system. These cause directly opposite effects to each other, so that the aggressive cat's pupils are constricted by the action of the parasympathetic nerves, while the defensive, fearful cat's pupils are dilated by the action of the sympathetic nerves. The relaxed cat's normal pupil size is due to the balancing of the two types of input. The nerve endings of the sympathetic nervous system release noradrenalin to the muscles to cause their effect: this is also a hormone released by the adrenal gland, so the nervous and hormonal systems are doubly linked and the hormone backs up the nerve transmission.

Alert at night, ready for fight or flight

Stress Hormones

The adrenal gland is particularly important in mediating the body's response to stress. In addition to producing adrenalin and noradrenalin, it produces cortisone, a steroid hormone that modifies the cat's metabolism in times of mental stress, such as territorial disputes, growing cat density and confinement, and physical stress. Consequently, measuring the levels of the adrenocorticotrophic hormone (ACTH) and cortisone in the cat's urine places a numerical value on the stress and emotional activity that a cat is experiencing at the time.

Sexual Activity

The trophic hormones from the pituitary gland control much of the sexual activity and timing. The raking of the male's penile barbs inside the queen's vagina (see page 55) is detected by the sensory nerves. The reflex message is conducted to the brain and hypothalamus, which causes the release of luteinizing hormone from the pituitary, which in turn causes the final maturing of the ovarian follicles and release of the ova.

The Cat's Memory

The dismissive attitude of Descartes and his forebears towards the ability of animals to think, in whatever form, is behind many people's ill-held belief that animals don't have memories.

Cats, like dogs, show classic Pavlovian conditioning when they suddenly appear at the exact time that you normally feed them. Pavlov trained his dogs to salivate when a bell was rung, by ringing a bell when he fed them. Then, when the bell was rung alone, they still arrived for food and so demonstrated the connection. When cats anticipate your feeding time (almost to the minute if you do it regularly enough), they are demonstrating a connection with a time rather than a bell. In Britain and the US, when the clocks are set back one hour with the changing season, cats all over the country are missing out on their meals! They may watch for clues, but where the situation does not provide them, then the event shows more than just a connection: it shows an accurate awareness of time and a memory.

When you regularly return home at a certain time, your cat accurately anticipates your arrival without a watch

Stress in Territoriality...

A cat feels confident in its territory if it does not detect intrusion. In areas where it is challenged its heart rate increases and it becomes stressed, and its autonomic nervous system and stress hormones make it ready to react. It will spray, urinate, leave droppings exposed and become ready to retaliate against the intruder. If the other cat retires, the resident cat regains its confidence, and stress disappears. Even if the resident concedes it, at least, is able to retreat.

When cats are confined their territories are not only restricted, but they are confined by solid walls and retreat options are reduced. If they are subject to disturbance their stress levels rise, and what we perceive as behaviour problems occur: spraying, urinating, 'missing the litter' and aggression. Yet these are the cats'

OUTSIDE

Above: A young cat entering into its owner's garden, but into an existing layout of cat territories in which currently there is no gap for the newcomer. Whenever a newcomer arrives there is a period of readjustment of territorial boundaries

Left: A neighbouring cat watches cautiously from the security of its resting spot. It is reassured by its own scent which repeated use of its lair has given it. Each cat's territorial confidence is dependent on its or its group's scent predominating

Right: The stress hormone levels of the cat into whose existing territory the newcomer is walking rise in response to the threat from the incursion. In response to a package of adrenalin, testosterone, noradrenalin and cortisone it becomes alert, its ears begin to swivel, its pupils constrict, its fur fluffs up and it raises its body profile in a threatening posture. Once the dispute is over hormone levels return towards normal

...and Confinement

INSIDE

Left: In confined cats, with only a limited area determined by walls, retreat is restricted and consequently stress hormone levels are more usually elevated. Hence stress-related behaviour is more frequently encountered with confinement. One escape that cats can have from other cats and stress is to climb up

appropriate responses to territorial stress.

When new adult cats are introduced into confined multi-cat households there is a prolonged period of adjustment during which time cats will furtively watch each other and there will be aggressive interactions. Things often take longer to settle down than owners hope for; it is a territorially stressing event. However, if introductions are managed, behaviour will normally have shifted to be less wary and more socially interactive within a year, with even mutual grooming occurring. However, when other stressors impinge conditions may remain uneasy with 'problems'.

Above: Mealtime is approaching, with consequent increased crowding of cats in a multi-cat household, so behavioural anxiety and aggression levels respond to the rising stress hormone levels. A dominant male corners a female who having no retreat is forced to retaliate defensively, and the male turns away. Indoors such moves are usually more restricted

Left: If normally outdoor cats are confined for a period of a few weeks their aggressive interactions become more frequent as time passes

Curiosity and the Cat

The expression 'curiosity killed the cat' reflects a long-held belief that of all animals it is the cat which will continually show curiosity – exploring, creeping into hidden places and tentatively tapping unusual objects with its paws to test their responsiveness. This behaviour is not without potential risks, as the proverb makes quite clear.

Young cats are extremely inquisitive. Although it sometimes leads them into difficulties, this appetite for discovery has a very practical purpose. The young cat must learn about the world – which is completely new to it – very quickly: it has about six months, maybe less, in which to prepare itself for life as an adult, and must learn to hunt and deal efficiently with prey, familiarize itself with its environment and learn to keep out of trouble.

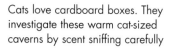

Cats love cardboard boxes. They investigate these warm cat-sized caverns by scent sniffing carefully

Cats will watch TV when an object moves at a speed similar to that of prey, and prey shapes and sounds can be perceived. Cats do not respond until their attention is caught by movements ten times faster than we would note, corresponding to prey movements

'Cat and Mouse'

Investigation of new objects certainly forms part of some cat play sequences and is particularly noticeable in post-weaned kittens around ten weeks of age. It is also part of their development to check out the state of small prey. During a hunting sequence and the dazing of prey that follows capture, the cat exhibits the necessary alertness and caution. For the small-prey hunting cat, continuous curiosity is essential (between naps!) as it must always be on the lookout for potential food.

The slightest small movement alerts a cat, while subsequent stillness from the prey invokes a tentative paw tap to test its state of awareness. Sudden movement from the prey brings about the resumption of the chase. This responsiveness to small movements is exploited by those who like to play 'cat and mouse' games with their cat, teasing them with a piece of string, or a commercially available toy on a thread and rod.

Hunting and Hunger

Paul Leyhausen demonstrated that in the cat the urge to catch goes beyond the immediate need for food. He released one mouse after another in front of a caged cat and found that it was ever ready to catch more. When it had a number in its mouth, and one under each forepaw, it *still* tried to catch more. Leyhausen suggested that there is a ranking in drives for survival, and hunting scored over hunger. For a feline hunter of small prey, this urge to hunt is vital for survival. Yet it produces behaviour that is often termed 'curiosity' by owners, such as the readiness with which a cat will run to the refrigerator (or wherever its food is kept) whenever it opens, showing interest even when it isn't hungry.

Right: Cats always have to be alert for foes as well as food, and keep careful watch on other cats as well as prey

Below: Our cars fascinate cats. They are initially wary, but our scents, useage, and an open door prove too much of an invitation for the curious cat

Common Behaviour Problems

Introduction

As we and cats are not the same species, it is remarkable that we can and do live together with so few difficulties. After all, there are enough problems when people live with people! However, since the 1980s there has been a massive increase in reported cat 'problems', or antisocial behaviour. Three factors have come into play at the same time as this noted increase in problems. The first is

the advent of cat behavioural practitioners, who not only give advice, helping owners to do something about their cat problems, but have been able to record the problems more rigorously than was done before. Secondly, across America, Britain and Europe there has been a huge increase in multi-cat households. Thirdly, and most significantly, there has been a trend – spreading first across the USA and then elsewhere – of confining cats inside houses or apartments.

The problems with reported cat behaviour are varied and include confinement stress, fouling, spraying, furniture damage, aggression, paddling and wool sucking, and bringing home prey. This section looks into the causes of these common difficulties and suggests solutions or tactics to ease them.

Animal behaviour counsellors use a range of techniques to counter cat behaviour that their owners find inappropriate, from hidden water pistols to advising medication. While the hierarchical nature of dogs allows dominance by an owner in dog training, the scolding of a cat is a show of aggression which is usually counter-productive. As cats behave to us as if we are members of a cat group, behaving aggressively towards your cat will weaken your bond, increase your cat's anxiety and decrease its confidence – all of which are likely to make any behavioural problems worse, not better.

Behaviourists do use active deterrents such as water pistols and tiny bean bags, but it is nearly always better to use passive techniques, such as those suggested in the Solutions sections in the following pages. In all circumstances, the causes of the problem should be examined and dealt with if possible.

Confinement

Confinement and the stress it can induce are major causes of behaviour problems in cats. Cats are extremely territorial, but their ranges are shrinking. In dense urban housing a female house cat may be limited to a range of 0.02ha (0.05 acre), yet many normal dwellings in which owners now live with their cats are only a tenth of that size.

Keeping additional cats in that confined space compounds the problem. Furthermore, as cats relate to us as if we are some form of cat, any increase in the human population in that space – including the arrival of a new baby or partner – will aggravate the situation. Owners using confinement as a way of protecting their cat from the outside world of traffic injury and disease, or who fear the cat may not return, have thus traded these dangers for confinement stress for their pet and problems for themselves. Plus in small featureless apartments cats can become bored. Ironically, the ready availability of cat litter has led to a huge increase in unnecessary confinement.

Confinement is not responsible for all behaviour problems, and indoor/outdoor cats also have difficulties, but these too are usually related to density factors. Confinement is not only a reduction of territory, but also a restriction in the cat's range of stimuli, hunting potential and escape options, which can also affect outdoor cats and cause them stress. Both indoor and outdoor cats are vulnerable to noise and aggression trauma. A stressed queen may respond by abandoning her kittens, engaging in excessive neonatal grooming or suffering reduced milk production. However, confinement stress commonly causes problems such as fouling, spraying, aggression, excessive grooming and eating disorders.

Confinement Stress

Stress causes physiological changes: the sympathetic autonomic nervous system responds by increasing heart rate, changing blood flow and releasing adrenalin, ready for 'fight' or 'flight'.

With territorial mammals like cats, an initial aggressive interaction can establish relative positions of dominance. Normally the dispute is settled by one withdrawing from the scene, while the other gains the territorial rights to the area. Where a number of cats are kept housebound, their territorial limits have been imposed on them. Any aggressive interactions which occur cannot therefore be resolved by the loser withdrawing from the other cat's area, and it will have to endure closer contact with the dominant cat than it would like.

Captivity stress like this can be very damaging and the prolonged biochemical effects may cause a whole range of pathological conditions, from fertility problems to heart failure. When cortisol levels remain high during stress, the allergic and inflammatory responses of the body are reduced, making the animal more vulnerable to infection.

Solutions

- Caring owners need to balance the relative risks of their local area carefully before choosing to confine their cats with the probability of stress.
- Confined cats should not just be allowed outside without a gradual introduction (preferably on a leash initially); the cat needs to assure itself of its range and way home.
- If cats remain confined then a great stress reducer is to give them the scope to climb up. Provide a cat gym and additional 'retreat shelving' around a room. (You put up shelving for plants and books, so why not your cat?) It also increases the range size. Consider making an extension via a cat flap to an external meshed conservatory.
- Giving access to a bedroom can reduce stress (see page 99).

Fouling Around the Home

Cats are often referred to by owners as naturally clean animals, by which they mean that they did not have to coach them to become house-trained. During early kittenhood, kittens only eliminate in response to their mother licking them in the appropriate area. In catteries, they need only a litter tray with its diggable surface and they will use it. However, at least 10 per cent of house cats will have a problem with fouling at some point in their lives, and it is the major problem reported to pet behaviourists. On the other hand, owners often confuse spraying with urination if they have not seen the event. Spraying (see page 124) is usually carried out against a vertical surface, while urination is normally on a flat surface. Urine spraying can be a problem with Siamese, while Persians are particularly prone to soiling.

Dirty Litter Trays

The most common cause of fouling around the house is failure to clean out and change the litter frequently enough. Cats vary in their fussiness over litter cleanliness. They normally bury their faeces, an activity connected with marking their territory, which in turn is related to confidence. Your cat may not be confident enough to use the litter tray where you have placed it. In addition, indoor/outdoor male cats may also be less at ease with eliminating indoors.

Environmental Conditions

Owners are more than twice as likely to have problems with inappropriate urination as with defecation. Indoor/outdoor cats are far less likely to present a problem, for they tend to both urinate and defecate towards the edge of their territory. However, both confidence and temperature can affect the distance from the door, and thus their range of latrine sites. Cold, and particularly intense frost and snow, can restrict the cats' practical access to digging, although at times some cats will use snow as if it were soil.

Cat Density

Cats can also become physically trapped in the house. Timid cats can sometimes feel trapped in by disturbances, and in multi-cat households with a cat flap exit one cat can intimidate another from its easy use. An increase in the number of cats outside, particularly aggressive ones, can reduce your cat's confidence about going outside. Similarly, additional cats or people inside the house can increase the density, reducing the cat's territorial confidence and leading to soiling.

Infection or Reduced Mobility

Problems like cystitis, which middle-aged female cats may develop at some time, can cause a cat suddenly to start fouling. If that is the case, easy access to cat litter and antibiotics should resolve the problem quickly. Elderly cats may also have mobility problems, which can affect their ability to eliminate in the appropriate place.

When fouling in the home has been caused by confinement stress (which is a major factor), the approach needs to be twofold. While the specific features of fouling should be attended to as outlined below, at the same time, as with the other stress-related problems, it is vital to tackle the underlying confinement stress (see page 121).

Solutions

- Take care with the siting of the litter tray so that your cat feels safe and confident when using it. A tray hood may provide the necessary sense of security that some cats require. Avoid placing the litter tray in close proximity to the cat's food bowl.
- Remember that in confining the cat you are not just reducing its range size, but also its choice of latrine marking sites, so it can be helpful to have more than one litter tray.
- The choice of litter can also be critical for many cats. Those that absorb and clump together to facilitate cleaning are best.
- When cleaning up after fouling, remember that cats tend to use certain latrine sites repeatedly due to the confidence gained from the scent of previous use. It is therefore vital to clean a fouled area in the home thoroughly

so as not to encourage reuse. Take care that the cleaning does not cause damage to fabrics, and avoid cleaners that may be toxic to cats, such as coal-tar products. The use of ammonia-based products is not recommended, for their smell can encourage urination. Sodium hypochlorite products or enzyme-based products made for the purpose can be most effective.

■ Having eradicated the smell, change the geography of the home by putting things in the way of the cat. This could be a sheet of aluminium foil, which some cats don't like walking on, or you can put ornaments or more furniture in the way so that the cat physically cannot reach the spot previously used.

■ You may feel irate when your favourite carpet is fouled, but it is likely that any punishment or show of aggression by you will simply increase the cat's anxiety and the likelihood of a recurrence. Instead, analyse why the problem has occurred and try to address the cause.

In multi-cat households fouling is most likely due to territorial stresses. The position of the litter tray becomes particularly important for cat confidence. Don't place it near the feeding area as some cats will then not use the litter; while fouled litter can put some cats off their food

Spraying

One of the major problems for most cat owners, sometimes even those who allow their cat freedom outside, is that of spraying, which makes up around 45 per cent of complaints about soiling. Yet such spraying is often not caused by their own cat but by an interloper from nearby. However, if you keep intact (uncastrated) male cats in your home for breeding, then the animals involved will probably be your own.

Spraying as a form of scent marking is carried out on a territorial basis by intact males. Consequently, castration normally reduces its frequency and pungency. However, neutered toms that are stressed by the arrival of new toms in their territorial domain may well spray, but it will lack the all-invasive pungency of that of an intact male. Even queens may deposit urine in a male-like way during their breeding period. To an intact tom, the odour of his own spray reassures him of his right to be there, and where other scents are detected he is likely to mark increasingly by fresh spraying. Many stud toms cannot mate unless their mating pen has been properly scented by spray.

Solutions

- While castration generally eliminates the problem of spraying, one in ten will remain a persistent sprayer. For these, treatment with progesterone is normally effective. Interestingly, studies have shown that neutering early does not make spraying any less likely.

- It is possible to dissuade a cat from spraying by employing behavioural techniques. Using a water pistol distantly and unseen when the cat is spraying can help to dissuade him. Aggressive irritability by owners is much less helpful.

- Try to analyse where and why your cat sprays. There are some surfaces which can almost *impel* a sprayer to spray. In the outside world, the size, shape and height of most car hub caps encourage spraying. Start looking at your home with this perspective and you will soon realise not only where such sites are, but how few they are. These can then be made less accessible with a change of local geography!

- In multi-cat households where spraying is a serious problem, identification of the perpetrator is not always easy. It is possible to 'label' the spray by having a veterinarian inject a suspect cat with a fluorescent marker dye. Over the twenty-four hours following injection the sprays can be looked at with ultraviolet light – the marked cat's spray will glow bright green.

Furniture Damage

Furniture damage normally occurs through cats clawing. Damage can occur when a cat runs up fabric, especially curtains, or if the claws are overlong and snag on chair fabric. There may be practical reasons for a cat wanting to claw around the home, and it is helpful to know why, when and where the cat does it.

The usual reason is a marking function. In the outside world cats will claw trees and fences for a range of reasons (see page 16). These also apply indoors. The marking function leaves a ragged visual signal which invites the cat to repeat its behaviour. Allowing the problem to arise and develop means it will probably get worse.

The nature of a surface affects the cat's choice. Hard, shiny materials will not accept the cat's claws, while unvarnished softwoods like pine are particularly favoured, as are padded fabrics. The legs of softwood tables and the vertical faces of upholstered chair arms are very commonly used.

In single-cat households, the cat will not mark much in this way in your absence, as this sort of damage may be a visual assurance signal. However, like most marking actions of domestic

cats, it is linked with confidence. When a frayed surface is read by other cats, its significance is conveyed by its position within the animal's territory. Yet the territorial size depends on the cat's confidence. Our presence lends confidence to our cats and they become more assured. Consequently, one common place that cats claw is the doormat at the entrance door to the home. As the cat comes in through a narrow entrance from outside, where it has had to keep alert to a range of threats of territorial intrusion, it enters not only the security of your mutual ranges, but finds that you are there – and presto, your cat claws the matting.

The anticipation of food given by you, with the security that conveys, means that kitchen tables are frequently clawed. For the same reason, rooms that you frequent are more likely to be claw marked than others.

Solutions

- A purpose-built scratching post is preferable to your prized furniture, and it needs to be positioned with regard to the cat's movements. This siting is critical. Cats often need encouragement to accept a scratching post.

Left: As claw marks attract the cat to re-mark, it is critical to deter the cat's action before marking becomes noticeable

Right: Once clawing has gone as far as this, you have left it too late

Playing with it with them by pulling a string over it or dragging a toy encourages its use.
- To dissuade a cat from damaging wooden furniture, try using a passive deterrent such as a strong-smelling polish. At the same time, remove the visual signal that is encouraging the cat towards that marking place – that is, remove the scratch marks.
- Covering furniture with a rug or aluminium foil, or placing things in the way, are both useful deterrents. If the cat does start to scratch, try gently transferring it to the scratching post sited in front of the marking spot.
- Cats have a tendency to scratch wooden stair-posts just inside the door that leads to the outside. To prevent this, wrap the post in aluminium foil and place an alternative 'scratching post' of a rush mat inside the door. Once the cat is using the mat, remove the foil and get rid of any visible fray or claw marks by refinishing the post's surface.

Aggression

1 Two house cats meet on the path of the resident cat.
The defending cat immediately 'fluffs up', as she stops
in the path of the potentially aggressive intruder

After fouling problems, 'aggression' in their pet is the second most common reason why cat owners consult an animal behaviourist. Yet of all the 'behaviour problems' it is the broadest in scope and the most ambiguous, for what is thought of as serious aggression by some owners would be considered merely a brief scrap or just a 'stand-off' by others.

Fortunately, full aggressive behaviour towards owners is not usually a major problem with cats, although some owners can receive quite bad lacerations from play that has developed into what an owner will perceive as an attack on their forearm.

There is also a broad spectrum of types of aggression, and these are part of the cat's natural

4 Her territory successfully defended, without coming to blows, she continues to stand her ground

5 The ex-aggressor retreats with a measured walk, and then looks back from a safe distance. Unlike confined cats, indoor/outdoor cats can more easily retreat as an option

The 'stand-off' (below): most aggressive encounters are settled without coming to blows

2 The intruder, initially confident, takes stock of the situation

3 The intruder sneaks an anxiety lick, while the defender checks behind her

repertoire of behavioural responses. Yet what may be fully appropriate for a cat's life, we may find inappropriate, concerning or even sometimes dangerous. It therefore becomes clear that aggression cannot automatically be characterised as something 'bad', for it depends upon appropriateness in a particular situation.

For an animal which survives as a lone individual for much of its life when living feral, aggression is an essential asset to draw upon in defending its territory against intruders. Nevertheless, feral cats are usually involved in fewer aggressive interactions than suburban house cats, which are living at higher densities. Owners of multi-cat households will see some form of aggression among their cats most frequently of all.

6 The defender walks confidently back along the path having been empowered by territorial ownership. Most territorial encounters are settled by postures and threats rather than developing into full fights (see pages 84–5)

Types of Aggression

The problems likely to cause most concern are male-to-male fighting, territorial aggression and fear aggression. However, in different situations these classifications can become blurred, and may occur together in a single conflict.

Maternal Aggression

Some types of aggression, such as the maternal aggression displayed by a mother cat defending her young against another, too inquisitive cat, are not only appropriate but, due to the context, hardly a problem.

Redirected Aggression

This type of aggression is certainly seen in multi-cat households and in areas where cats live at high density. It is most usually redirected at people during a visit to the vet. The anxious cat is held firmly and treated by the vet, but on its release from the source of danger the cat may lash out at its owner, redirecting its aggression. If this has occurred before, it can be anticipated, and with care avoided.

Threat Aggression

One of the problems associated with multi-cat households is a breakout of displays and threat aggression as food is about to be put down. Due to the density of the cats, some will be anxious about obtaining their share and almost invariably the same cat will turn upon another that it can intimidate easily, which again is usually the same cat each time. This type of aggression can be greatly reduced by feeding some of the cats in different areas.

Solutions

- To reduce threat aggression at meal times, feed your cats from separate bowls and in separate areas if possible.
- Neutering should significantly reduce male-to-male fighting. Persistent scrappers may benefit from treatment with a female hormone.
- Take care when introducing a new cat or kitten to the household. The simplest way is to bring in two kittens together, and they will grow up and behave as if they were litter mates. When introducing adult cats, allow their scents to meet before they do. Let one move around in the home in the absence of the other, then reverse the situation. You can extend this by having one in a basket in a room with the other, allowing them to become even more familiar, and then swapping them around. When you eventually allow the cats to meet, anticipate some skirmishes.

MALE-TO-MALE FIGHTING

Fighting between male cats can be a real problem if there is a dominant cat in the area. He will invariably be uncastrated, and will usually walk with a 'swagger' and a rolling gait, for the testosterone does produce distinct build differences, most noticeably a fuller-jowelled head. In nine cases out of ten, neutering will cause the fighting to cease, with over half the cats stopping straight away, while in others the involvement in fighting wanes over a matter of weeks. More usually, neutering of tom cats is performed routinely before they are sexually mature, and it appears that, again, one in ten will become fighters. For the remaining intransigent few, treatment with a female hormone, progestin, is effective in 75 per cent of cases.

The danger of redirected aggression

Above: The cat being stroked on the ground spots a territorial intruder

Below: It gains confidence from the handler's presence and height, and snarls at the intruder

Right: However, as it is being restrained by being held, the cat redirects its aggression towards the handler

Rough Play

Most of the time play with cats and ourselves, as between members of cat litters, is within recognised limits, with both sides being genuinely playful. We tend not to be too rough, mindful that the cat is smaller, and we are also cautious of the potential of claws. Cats are aware that we are much larger and more powerful. Adult cats of the same household, particularly ones that have been brought up together from a young age, are often playful with each other, employing chases and mock attacks. However, not

Play can suddenly become less fun for the owner when scratches are made through a covering shirt. Yet the cat's ear, eye and tail signals combined with the cat's fighting limb position show this is so for the cat as well

uncommonly these become more earnest than the householder would like, even if not the full-blown fights of antagonistic cats. Such boisterous play is common in confined cats where they have less opportunites to diffuse their activites than cats with access to the outside. Increased opportunities for play in the interior habitat can lessen the problem.

We could draw the line between ourselves and a cat and can implement it due to our size more easily than cats between themselves. In the behavioural sense our play with cats is 'less adult'.

Yet when a cat has been playing co-operatively with an owner, it may suddenly grab the wrist with its front paws and vigorously kick with both back legs simultaneously, or alternate individual back legs, and it may even attempt to bite the wrist. This is normal defensive fighting behaviour where a cat on its back rakes the other cat's underside with its back claws. (When using alternate feet even at the inhibited rate on our arm the usual rate of kick at 4 per second is faster than the more measured double kick at 3 per second.)

It understandably concerns owners, and is the most likely time that an owner will receive a scratched arm. The play has heightened the cat's mood, and the event will have occurred when the cat has rolled onto its back and the owner has attempted to tickle the cat's tummy. It is the area that is vulnerable during a full attack, and it may be that the contact with the stomach or your looming hand triggers the response. We feel the cat is behaving in a 'Jekyll and Hyde' manner, but to the cat it may seem that we have transformed from stroking in the same way as its mother's tongue used to; to suddenly threatening a full adult fight.

There are clearly variations between different cats, and there is a suggestion that there is a tolerance threshold of our behaviour, or trust, based on the kittenhood habituation period, so that habituated kittens are more trusting. However, I do find dependant cats less prone to the behaviour.

Solutions

- If you find your cat starting the sequence, pulling away will cause the cat to grip and kick harder. Go limp and the cat will stop. Distract it with your other hand, then remove your gripped hand and stand up. If you anticipate the cat's move, just stand up and avoid it.

Rubbing Armpits

In our armpits we have scent glands which produce smells that contain some sexual aroma. However clean you are, there will be a residual amount there. It should not surprise us that, with their sensitive noses, cats will notice such things about us – when it has been found that people can tell the sex of another person by the smell of their sweat alone.

Cats are attracted by our aroma and this may lead to armpit rubbing. Your cat starts sniffing, then burrows its chin and head into your armpit, rubbing its chin and nose right up into the area of sweat production. If this is tolerated, the cat will repeatedly burrow and rub, drooling with its mouth partly open, and may even grip your clothing briefly with its teeth. If your shoulders are uncovered, the cat may similarly tug at your armpit hair and even lick your sweat. This behaviour is virtually identical with that of a cat sniffing and rubbing on the catnip plant (see page 43). If stroked even lightly during this time, the cat is likely to raise its rump and tail readily in a sexually aroused fashion. If not interrupted by your

hysteria, your cat is likely to go on for around a minute and a half before becoming satiated. The similarity of the cat's behaviour with armpits and catnip is due to a common sexual odour in both. It has been suggested that the cat is also seeking the lanolin 'fur' smell.

The attractiveness of the sexual aroma of the axilla and genital areas of worn clothing is probably the reason why cats like to lie, sleep and paddle on their owner's discarded clothes. This is also partly due to the comfort of the soft clothing and the reassurance of the owner's unique smell. The reassurance factor appears to be paramount, as it is dependent cats that are most attracted to warm discarded clothing.

Solutions

- As you are much taller than your cat, you can easily stop this behaviour as it happens, or even prevent it, by simply standing up. However, you may find friends begging you to allow it to continue for its amusement value to them!

Cats that react strongly to catnip are those most likely to carry out armpit rubbing

Paddling and Wool Sucking

Paddling

When you are sitting down, perhaps watching television, your cat may sit on your lap and start purring and paddling on your chest. Some people worry about this behaviour, but it is quite normal in house cats, for your cat is being 'juvenile' again, reliving its kittenhood and treating you as its mother. When it was very young, foot-paddling either side of its mother's nipple helped to stimulate milk flow. Our larger size and warmth make our adult cats feel like kittens beside their mothers. If you are armoured with stout clothing you may be fine, but if not and the cat's claws are sharp, you may not be feeling as contented as the cat! How often a cat paddles is usually a reflection of its character and how it responds to you. The less confident, dependent cat is far more likely to spend time puncturing your shoulder than the outgoing, laid-back cat.

Wool Sucking and Eating

If you are wearing a jumper and your cat starts paddling and purring, you may be protected by the texture of the garment. However, the lanolin smell from the wool may induce the cat to drool and suck at it. You may need to curb this habit, particularly in cats who drool so much that the woollen item becomes a wet mess. There does seem to be a strong genetic predisposition to wool sucking, with some breeds being determined wool suckers/eaters. This is most common in Siamese and Burmese, who can both behave quite obsessively in this way. Some cats will also eat cotton or synthetic fabrics, but the majority start with wool.

Some cats will eat other strange objects, such as elastic bands and paper. I once found a feral

Cats taken early from their mothers, before being fully weaned, tend to knead or paddle their owners more in adult life

cat that had eaten a small metal box – whole! The causes for this odd behaviour will not be the same in every case. For example, an adult cat developed the habit of chewing and shredding paper from having been provided with newspaper bedding in its box. It then extended this habit to rather more significant sheets of paper in the house.

Stress and boredom can cause a cat to react by turning to paddling or wool sucking, particularly if it is confined inside its owner's home. Anxious, dependent cats may also develop such behaviour.

Solutions

- One practical answer to painful paddling is to trim the cat's claws to blunt them.
- Another obvious answer is not to collude with this behaviour and simply stand up whenever it occurs.
- Wool sucking can be prevented by acting in the same way as a mother cat when she prevents suckling. As the kittens are approaching the time of weaning and she is cutting down on their milk, she will lie over on her nipples to avoid being continually pummelled. You too can move aside or away.
- Aim to give more access to the outside, especially for a confined cat.
- Try playing more with your cat. More time and reassurance can be a short-term help for an over-dependent cat, and even leaving material around with your odour on it can be of value. But for the over-dependence itself, you should look for a more long-term solution.
- Using nasty-tasting substances seems to have only limited value in preventing wool sucking.
- Reassess your cat's diet. It is noticeable that regular prey-hunting outdoor cats are less likely to be obsessive eaters of strange materials, and offering your cat interesting and naturally textured food may be helpful.

Ear Lobe Sucking

Some cats develop the habit of sucking their owner's ear lobes. This is an interesting variant on wool sucking, which is itself an echo of suckling.

If ear-lobe sucking becomes a serious habit, causing owners to lose sleep, it may be necessary to shut the cat out of the bedroom

When ear-lobe sucking occurs, the person is usually lying down in bed and may even be asleep. Kittens that are weaned too early, or have had a restricted milk supply from their mother, may continue to suck at parts of the bodies of their litter mates. They are also likely to be excessive paddlers and over-dependent.

Solutions

- Standing up or moving away each time are quite effective in cutting down on this behaviour.
- Taste aversion can also be tried.

Excessive Grooming

Stress can cause a cat to fail to keep itself in good condition, and overgrooming is a much more common response than undergrooming. The stresses of confinement in particular can cause a problem with overgrooming.

Fleas If the cat develops a sudden urgency in grooming, it is always sensible to check for fleas. Comb the cat on white paper – black specks of flea droppings will be seen if the parasites are present. A cat's attack on fleas can itself provoke overgrooming, leading to skin irritation, which in turn can cause the cat to continue to overgroom to the point of creating near baldness in the affected area.

In addition to the problems fleas can cause cats through overgrooming, when flea infestations take hold in part of a building that a cat normally uses, the cat will reduce its time in those areas. At such times, which can coincide with hot weather, the cat may even try to avoid walking on the carpet in infested areas, and will negotiate its way about up on the furniture.

Confinement Overgrooming occurs more commonly in confined cats than in those given access to the range of interests provided by the outside world. It can be due to boredom and is often found in anxious cats. Although overgrooming may arise in any cat, Siamese and Burmese are particularly prone to it, as are nervy Abyssinians.

Signs of overgrooming can be seen on the lower back, abdomen and inner thighs. Loss of hair behind the ears or on part of the back caused by repeated, vigorous scratching with the back feet usually arises from a reaction to skin parasites.

This cat is overweight, and overgrooms. It has denuded part of its underside by obsessive grooming

Solutions

- Where fleas are the problem, the first priority is to remove them. This normally involves the use of a proprietary pesticide sold specifically for the purpose. Depending on the product, this is applied as either a powder or a spray, given in droplet form by mouth, or absorbed through the skin from the back of the neck. However, you will need to treat the house at the same time as the cat, as most fleas are hit-and-run merchants and will lurk in your carpets until they are ready to strike again. It may be necessary to repeat the treatment: check the instructions given with the pesticide or ask your vet.
- Boredom from confinement can be alleviated by giving your cat more stimulation. This could take the form of providing an 'entertainment centre' based around a cat gym.
- If the problem is density stress, whether caused by other cats or by people, the exact cause needs to be identified and removed if at all possible.
- Overgrooming can also arise when a dependent cat is left on its own for longer periods than previously. In this case, if more time cannot be given to the cat, the closeness of the owner can be mimicked – for example, by leaving around recently worn clothing with the owner's scent on it – to provide reassurance.

Bringing Home Prey

Even the most ardent cat lover can become concerned when their cat brings home prey. However, the cat is a carnivore and it is not amoral for it to catch prey. Some owners seem very concerned over the capture of one wren, while they accept or are even pleased at the dispatching of ten rodents. And although most owners dislike their cats bringing home prey, this is made worse if the prey escapes. While they may wish the prey had been left alive outside, they don't want it left alive inside!

In reality, studies from all around the world have found that cats catch relatively few birds compared to small mammals. In a study of ringed birds caught by cats in gardens, Chris Mead of the British Trust for Ornithology found that in Britain cats were not having a harmful effect on bird populations. Further, we do not just support the cat population, but through our direct feeding of birds, and the provision of good habitat and nesting sites in our gardens and buildings, we are artificially sustaining a much higher bird population than can be supported out in the harsher environment of the countryside.

You may be reassured by this but still wish to reduce the effect of your cat on your garden wildlife. If you live in a town, as most people do, you have already reduced your cat's ability to catch much as ranges are smaller, for example, I found the average annual catch of the average London cat to be two items instead of the fourteen of a village cat. In addition, if you choose your cat as a kitten from a litter that was born and reared with the mother confined (so that she could not introduce the kittens to prey during the sensitive period approaching weaning), then your kitten is far less likely to develop into a masterful hunter.

As most cats live in towns the annual figure of catches in Britain is just 28 million, a fraction of the inflated figures sometimes suggested

Solutions

- You are extremely unlikely to be there at the capture, but if your cat has returned to your garden and is carrying out a capture and release sequence of behaviour (see pages 48–9), the judicious use of a water pistol can sometimes be effective. The cat must be unaware that it is you doing the firing. The disadvantage of using this technique in the house is that you will probably end up with an escaped mouse hiding behind your furniture.

- If prey is released into the house, then the Tabor Welly Technique can be helpful. As mice and other small mammals tend to run along the edges of rooms, if you position a gumboot on its side along the wall ahead of fleeing prey, with the open end towards the prey, it will gratefully dive into the security of the dark tunnel. You can then quietly take boot and animal outside to release it.

- To reduce the probability of your cat capturing birds in the garden, remember that cats are much more successful at catching birds on the ground than in trees or on bird tables, so feed birds at high levels. You can make your bird table less cat friendly by not using a rustic pole which is relatively easy to climb. An anti-climb cone, suspended tables or tables on metal poles can be helpful, but the usual design with a large platform is also quite difficult for a cat to scramble up. Time is the key factor, for the longer it takes a cat to get up, the more time the bird will have to spot it and fly off.

Eating House Plants

Carnivorous cats do not have the teeth to chew and grind vegetation. One look at your cat in the garden as it awkwardly chews off a blade of grass with its side teeth and then gulps it down unmasticated will convince you that it is not a herbivore! Nonetheless, cats do eat some vegetation, mainly grass and leaves. This practice may provide roughage to aid digestion. When cats vomit up hairballs, grass is usually entwined, and it may assist in removing the discomforting mass.

Whatever the reason, if you have confined cats it is sensible to provide a pot of grass in the house, to reduce the damage to potted plants. Avoid keeping cats confined with access to any plants that are poisonous to them. A number of house plants have been connected with the poisoning of cats, including poinsettia, philodendron, ivy, dieffenbachia, azalea, Christmas cherry, and even mistletoe and holly common among Christmas decorations. When vases of cut flowers such as sweet peas, delphiniums and lupins are added to the list, an owner's anxiety escalates! The danger from house plants is not

The narrow leaves of some house plants can seem temptingly like grass!

usually as extreme as from toxins such as antifreeze, but for some cats, particularly confined ones, it can be serious. Usually there is a localised ulceration and irritation in the mouth, sometimes accompanied by gastric disturbance.

Solutions

- If the cat is caught in the act it can be dissuaded by a water pistol, or a small bean bag, aimed unseen by the owner. As these active deterrents can misfire, and also require your constant presence, it is probably more effective to try passive techniques.
- Position house plants so that they are less accessible, but beware of cats climbing onto shelves. Placing the plant on a sheet of aluminium foil may not look attractive, but it may be effective in deterring the determined nibbler.
- Some cats repeatedly dig around house plants in large pots, trying to use the compost as litter. Again, blocking their access is effective, but so too is placing mothballs in a muslin bag on the surface of the soil, as cats dislike the smell. It can also prevent them from damaging the leaves. Vinegar and other unpleasant-smelling substances can also be used.

Climbing Up

Cats have an insatiable and perfectly understandable desire to 'get up' onto higher levels. They are, after all, tree-climbing mammals that equate safety with being up high, well out of trouble, in a place from which they can survey the world around them. It is this that drives cats to leap up onto furniture in preference to walking across the floor.

The kitchen is a dangerous place for the cat who wants to 'get up'. It is essential, of course, that cats do not get onto the hotplates of cooking stoves, and it is far from ideal to have them walking on food preparation surfaces. However, cutting up and dishing out their food on these surfaces does mean that they will be looked on favourably by cats. Most owners will just scream 'No!', which is of limited effectiveness, for yelling at your cat can

Cats naturally wish to climb up for their safety and to investigate, but that is not a good idea in the kitchen

weaken the bond you have developed with it. In addition, although the cat may appreciate the fact that you do not want it there and will humour you while you are about, it may see no reason for the same restriction when you are not. Usually, adult cats will be more sensible than kittens with heat.

Solutions

- Try to alter the local geography of the home to make danger sites less accessible – for example, by leaving objects on surfaces. This becomes increasingly necessary if you have a number of housebound cats, for due to territorial tightness they will be particularly keen to get up onto spots which they feel are secure.
- Providing some new surfaces, such as cat gyms, and allowing cats up onto non-damaging surfaces like internal windowsills is very helpful. Do not worry that providing alternative sites that the cat can get up onto will confuse it, for the very concept of not climbing is completely alien to a cat.
- The limited use of a water pistol, with the water arriving out of the blue and apparently not from you, can be invaluable for deterring cats who appear to be oblivious to danger. The cat then connects that particular spot with the very real danger of a water squirt.

A cat gym provides the cat with its own safe place where it can climb up, play and have a stress-relieving retreat

Feline Obesity

It was once possible to say that cats rarely have feeding problems – there have always been far greater difficulties with dogs. However, recent surveys in the United States have revealed that up to one-third of household cats are now overweight, mirroring the increase in obesity in the human population. This, of course, is also happening in Britain and Europe, albeit on a much-reduced scale, so it may not be long before we see a significant increase in obesity in the cat population in these countries too.

Presenting a cat with over-large portions of food is a major cause of feline obesity. Another contributing factor is the huge increase in confinement of cats which has been observed in the United States. This results in a dangerous reduction in activity levels due to the restricted area, and an increase in boredom.

Owners who go out to work often leave out

An adult male unneutered breed cat which, despite its massive head and full cheeks, is not overweight. Remember to take your cat's breed into account when assessing whether or not it has a weight problem

more food for their cat than they should, in an attempt to compensate for the guilt they feel at leaving their pet. However, extra-large portions are no way to express care and affection – quite the reverse – and portion size should reflect the cat's real needs.

Part of the problem is that, in response to market forces, cat-food manufacturers have aimed to produce highly palatable foods, which encourage cats to overeat if given the chance.

The increase in multi-cat households in recent years means that tensions can run high at feeding time, accentuating the tendency of some cats to gorge, while others will eat less than they need.

Unlike the cat on the left, this one *is* overweight: when you run a multi-cat household it can be difficult to monitor how much or how little each cat is eating. However this cat has not yet become obese, unlike the cat on page 134 where the condition is causing the animal problems

Solutions

- Do not offer your cat portions of food which are larger than it requires to maintain its correct body weight.
- If obesity is the result of inactivity caused by confinement, the most effective cure is to give the cat normal access to the outside world once again, allowing it to set its own range. If this is not possible, you should compensate by increasing the time you spend playing with your cat, perhaps using a cat 'gym' to raise the levels of activity.

In multi-cat households feeding cats closely together can lead to some cats overeating and others undereating. Feeding in different rooms and not leaving a 'gorger' with access to other cats' uncovered food can help

Finicky Feeders

Cats are notoriously finicky feeders and many households experience a battle of wills over meals, with the owner putting down a particular type of food for the cat and the cat refusing to touch it. Owners will sometimes do this for days before finally giving in to the cat!

Where food is concerned, the cats' iron will is amply demonstrated by their attachment to particular brands of catfood, which has given pet food manufacturers an advantage over supermarkets trying to introduce their own brands. Cats have reportedly starved to death rather than eat what they find unpalatable. The basis for this stems from the narrow food preference that they learned as kittens: whatever mother brought in when they were kittens they accept, and what they ate then they have a preference for now.

Cats prefer meat, and particular types. However, when they are sick, and especially when suffering from an illness that affects the nose, cats can go off their food as their initial response to it is through smell.

Solutions

- Do not serve a cat its food near a litter tray, which can put it off.
- Do not serve food straight from the refrigerator, as cats do not like cold food. It is not just the temperature on the tongue, but the fact that warmer food releases more scent. Cats have evolved to eat food at body temperature (37°C/99°F) and that remains their preference. However, as their interest does not increase markedly between 25°C and 45°C (77°F and 112°F), in all but the most resistant cat there is little to be gained by raising the temperature beyond 25°C (77°F).
- Food left out uncovered for too long can age, dry out and even attract flies, and the cat will avoid it.
- Cats can sometimes be weaned onto new foods by mixing gradually increasing amounts of the new food with accepted food.

UNDEREATING

If your cat suddenly loses weight, consult your vet immediately, as there may be a kidney problem. If you have an elderly cat, keep a close eye on its condition when stroking along the spine, and watch out for weight loss accompanied by drooling, which can indicate gum infection and tooth loss.

In addition, if the cat's nose is congested as the result of an infection, it may lose all interest in its food.

This cat, almost twenty years old, shows weight loss due to kidney damage; a common problem in older cats

Roaming

This term covers a variety of behaviour patterns: cats that stray completely from their homes; cats that stay out for a number of nights; cats that have large ranges; and cats that just visit other houses. Straying completely commonly occurs when the owners move home and do not keep their cat indoors long enough for it to develop new territorial attachment before they allow it out. In some circumstances, the cat may find its way back to its old home. Incredibly, some owners make no allowance for their cat's territorial attachment; for example, in one case I was called to the owner had merely put the cat down on arrival in the new home, with doors open, workmen banging about, and removal men carrying furniture around. Needless to say, the cat disappeared within seconds, and its lady owner was distraught for weeks.

Straying can also occur when too many cats are put into one home, and one or more cats will leave to escape the stress. The same situation can cause some male cats suddenly to assume massively larger ranges than they held formely, and similarly greater than the normal size for that area. In contrast, a minority of males patrol a disproportionately large range – up to four times the normal size for the local area – throughout most of their lives. I documented the case of a male Siamese and found that its father had displayed similar behaviour, which therefore may have been inherited, genetically or even by being learned. Both cats could also be aggressive towards people, and assertive in demeanour.

Some male cats that have large ranges may stay out for several nights at a time. Some will be hunting, but others may be 'visiting'. Such cats may have a number of other 'homes' where they are fed, and in extreme situations more than one household may believe the cat belongs to them!

Solutions

- When you move house, it is essential to keep your cat safely confined to the new home for at least a week before giving it access to the outside world. Ideally, take it out on a long lead harness while it establishes a map of the immediate garden area, and keep the door open for its return. Increasingly confident behaviour will indicate when the cat is ready to go 'solo'.
- Avoid high stress levels that might make a cat leave home; some cats cannot cope with multi-cat households, or continual disturbance from building work and the like.
- Neutering male cats will generally reduce both their range size and potential for wandering.
- A cat wearing a name tag and collar, or which has an identification chip, is less likely to be mistaken for a stray in need of a home by well-meaning near-neighbours or welfare organisations.
- If you find your cat is also being fed by well-meaning neighbours, a tactful conversation will often remedy the situation.

Further Reading

Beaver, Bonnie (1992) *Feline Behaviour: A Guide for Veterinarians* (W. B. Saunders)

Borchelt, Peter & Voith, Victoria (1980) 'Social Behaviour of Domestic Cats' *Compendium Sm. Animal: Pract Vet.* 637–646

Bradshaw, John (1992) *The Behaviour of the Domestic Cat* (CAB International)

Caro, T. M. (1981) 'Predatory Behaviour and Social Play in Kittens', *Behaviour*, 76, 1–24

Eisner, T. (1964) 'The repelling of insects by Nepeta', *Science*, NY 146, 1318–1320

Fogle, Bruce (1991) *The Cat's Mind* (Pelham Books)

Frazer Sissom, D., Rice, D. & Peters, G. (1991) *How Cats Purr*, J. Zool., Lond. 223, 67–78

Hart, B. (1976) 'The Role of Grooming Activity', *Feline Pract*, 6:14

Hart, B. & Barrett, R. (1973) 'Effects of Castration on fighting, roaming and urine spraying in adult male cats', *J. Amer Vet. Assoc.* 163, 290

Hart, B. and Hart, L. (1985) *Canine & Feline Behavioural Therapy* (Lea & Febiger)

Johnson-Ory Gene (1991) *Getting to Know the Bengal Cat*

Karsh, Eileen (1983) 'The Effects of Early Handling on the Development of Social Bonds between Cats & People', in *New Perspectives on our Lives with Companion Animals*, ed. A. Katcher & A. Beck (University of Pennsylvania Press)

Kiley-Worthington, M. (1984) 'Animal Language? Vocal communication of some ungulates, canids and felids', *Acta Zool. Fenn.* 171, 83–88

Leyhausen, Paul (1979) *Cat Behaviour, The Predatory & Social Behaviour of Domestic & Wild Cats* (Garland STPM Press)

Mead, Chris (1987) 'Ringed Birds Killed by Cats' *Mammal Rev.* 12:4, 183–186

Moelk, Mildred (1944) 'Vocalising in the House Cat', *Amer. J. Psychol.* 57, 84–205

Morris, Desmond (1996) *Cat World: A Feline Encyclopedia* (Ebury Press)

Natoli, Eugenia (1985) 'Spacing patterns in a colony of urban stray cats in the historic centre of Rome', *App. Animal Ethol.*, 14, 289–304

Robinson, R. & Cox, H. (1970) 'Reproductive Performance in a cat colony over a 10 year period', *Lab. Animals*, 4, 99–112

Schaller, George (1972) *The Serengeti Lion: a study of predator-prey relations* (Univ. of Chicago Press)

Seidensticker, John & Lumpkin Susan (1991) *Great Cats* (Merehurst)

Simpson, Francis (1903) *The Book of the Cat* (Cassell)

Sinrod, Barry (1993) *Do you do it when your pet's in the room?* (Fawcett Columbia)

Tabor, Roger (1981) General Biology of Feral Cats in *The Ecology and Control of Feral Cats* (Universities Federation for Animal Welfare)

Tabor, Roger (1983) *The Wild Life of the Domestic Cat* (Arrow)

Tabor, Roger (1989) 'The Changing Life of Feral Cats at Home and Abroad', *Zool. J. Linnean Soc*, 95: 151–161

Tabor, Roger (1991) *Cats: The Rise of the Cat* (BBC Books)

Tabor, Roger (1995) *Understanding Cats* (David & Charles)

Turner, Dennis & Bateson, Patrick (1988) *The Domestic Cat* (Cambridge University Press)

Acknowledgments

This book is the companion volume to *Understanding Cats*, and, like it, arose from over two decades of my working as a cat biologist and behaviourist. For my research and for television I have followed the cat through many countries round the world, where I have been helped by innumerable kind people. I cannot possibly hope to thank everyone who has been involved, and I trust those that I do not mention by name will nonetheless know that they have my thanks.

Thanks are due to John Bowe, Colin Tennant and colleagues at Bowe-Tennant Productions, Dick Meadows, Betty Bealey and colleagues at the BBC, Johnny Morris, Desmond Morris, University of East London, Universities Federation for Animal Welfare, The Cats' Protection League HQ staff, Olive Hammond and Julie Kircher at White Motley CPL Shelter, Chitwan National Park (Nepal), Doris Westwood and the Fitzroy Square Frontagers and Gardens Committee, Debbie Rijnders and Ineke Aldendorff of Stichting De Zwaerfkat (Holland), London Zoo, Colchester Zoo, Tony Cook of The Wildfowl Trust, Becky Robinson, Louise Holton and colleagues of Alley Cat Allies, Andrew Mugford of Cressing Temple, ECC, Helena Sanders and Venice DINGO, TICA (New York and Los Angeles), Mike Jackson, and Mary Wyatt. Particular thanks for his co-operation and patience are due to the veterinary surgeon Alan Hatch. Further thanks are due to Pat Simon, Diane Slater, Anne Bailey, Jean Murchison, Mrs A Wright, Jan Cudlip, Bernice Mead, Barbara Castle, Dawn Gulliver, Mrs G. Day, Mr & Mrs D. Heath, Ed & Malee Rose, Hank & D.D. Tyler, Rosie Alger & Barrie Street, Mira Bar-Hillel & Geoffrey Addison, Steve & Margaret Cuthbert, Bob, Vally & Charlotte Hudson, Phillipa Spalding, Michael Harding, Robin & Georgie Kiashek, Caitlin Carr, Merryl Wilkinson, Sue Sanderman, Rachel & the Cooke family, Lord & Lady Fisher, Solvig Pfluger, Ann Baker, Joan Hodge, Charles Llewellyn, Norman Collus, Janet, John & Viki Collins, Jean Renney, and Ken & Betty Tabor.

I wish to express my sincere appreciation of all cats and cat owners who have assisted my rsearchers, giving rise to this book, particularly my own cats, Jeremy, Tabitha & Leroy.

Thanks are due to Shona Drury, Angela Weatherley & Mary Redman for help with the manuscript. Finally it gives me especial pleasure to record my particular thanks to Liz Artindale, without whose hard work and considerable support this book would have taken far longer to prepare, and I am grateful for the fine addition of her photographs to stand alongside mine.

ROGER TABOR, 1997

Index

Page numbers in *italic* indicate illustrations